Student Course Guide

for

Nutrition Pathways

Eighth Edition

MARIE YOST MANESS, Ph.D.
Content Specialist
Dallas County Community College District

Student Course Guide revised by
COLLEEN LOVELAND
Dallas County Community College District

For use with the thirteenth edition of *Understanding Nutrition*
by Ellie Whitney and Sharon Rady Rolfes.

Produced by:

Dallas TeleLearning

Dallas County Community College District
in association with

Australia • Brazil • Japan • Korea • Mexico • Singapore • Spain • United Kingdom • United States

Dallas TeleLearning
9596 Walnut Street
Dallas, Texas 75243

http://telelearning.dcccd.edu
1-866-347-8576

ISBN 13: 978-1-133-60448-8
ISBN 10: 1-133-60448-X

Wadsworth, Cengage Learning
20 Davis Drive
Belmont, CA 94002-3098
USA

Cengage Learning is a leading provider of customized learning solutions with office locations around the globe, including Singapore, the United Kingdom, Australia, Mexico, Brazil, and Japan. Locate your local office at:
www.cengage.com/global

Cengage Learning products are represented in Canada by Nelson Education, Ltd.

To learn more about Wadsworth, visit
www.cengage.com/wadsworth

Purchase any of our products at your local college store or at our preferred online store
www.CengageBrain.com

Printed in the United States of America
1 2 3 4 5 6 7 16 15 14 13 12

Contents

To the Student

Nutrition—there are so many questions and so much confusion surrounding nutrition. Are there really "good" and "bad" foods? Should you take dietary supplements and, if so, which ones? What is the best program for weight loss? How does exercise affect longevity? What exactly is metabolism? Is there really a connection between health status and nutrition choices? These and other questions will be answered as you explore *Nutrition Pathways*.

Throughout this course, you will learn to separate nutrition fact from fiction using the correct, scientifically based information. In this introductory course on human nutrition, you will receive a sampling of basic nutrition concepts, any one of which could be its own course! Although this distance learning course is designed to satisfy basic nutrition course requirements for college students entering allied health programs, it provides practical and interesting nutrition information to nonmajors as well. With *Nutrition Pathways*, you will discover that the nutrition and lifestyle choices you make can either positively or negatively affect your health and well-being. Through the video lessons that profile real people in real situations, outstanding experts and professionals associated with nutrition reveal to you just how the choices we make influence us every day.

It is our hope that you will be able to use the knowledge you have gained to improve the quality of your life and the lives of your family. So follow *Nutrition Pathways* to a destination that is healthier and more successful!

—Marie Yost Maness, Ph.D.
Nutrition Content Specialist

Acknowledgments

I wish to thank my husband, Ed, my daughter, Tracy, and my friend, Avis, who have been my support system throughout this project. I also wish to express my sincerest appreciation to several people who worked diligently and professionally to put together an exceptional educational product and gave me one of the best experiences of my professional life.

First, I want to thank Pamela Kettle, producer/director, for her creativity, strength of character, and willingness to learn and understand the course content by allowing herself to be a subject in the Pathways segments of the video lessons. I also thank Nora Coto Busby, instructional designer, who helped me learn how important it is to express myself clearly for the benefit of students who will be taking this course. After two years of close work, long hours, and much laughter, I consider both of these women to be special friends of mine. To Paul Bosner, director of production, I owe thanks for his guidance and questions that helped us form the basis of the project and for his suggestion that I pursue my Ph.D. in nutrition using the development of this course as the topic of my dissertation.

There would have been no interviews, schedules, or shoots without production assistants Debra Brown and Nicole Rambo. The video lessons would not have been made without the skill of the audio and video crews nor would they have been brought to life, depicting a dramatic story, without the expertise of editors Michael Coleman and Vicki Gratz. Thanks also to Betsy Turner who kept in contact with our publisher to keep us on schedule and to Lannie Waggy who kept the student course guide moving along. Other people supportive of this project were Pamela K. Quinn, provost of the R. Jan LeCroy Center for Educational Telecommunications, and the marketing staff.

My special thanks go to the best advisory committees a content specialist could have. They provided me with constant feedback on every aspect of course development from the formative stages through the final details of the telecourse. They were invaluable in helping *Nutrition Pathways* become a course that would bring pride to any educator.

My appreciation and thanks to Evelyn J. Wong, telecommunications information specialist, for her assistance with the subsequent editions.

Course Organization

Nutrition Pathways is designed as a comprehensive learning package consisting of four elements: student course guide, textbook, video programs, and interactive activities.

STUDENT COURSE GUIDE

The student course guide acts as your daily instructor. If you follow the Course Guidelines carefully, you should successfully complete all of the requirements for this course. (See the section entitled "Course Guidelines.")

Maness, Marie Yost. *Student Course Guide for Nutrition Pathways: Introduction to Nutrition.* 8th ed. Belmont, Calif.: Wadsworth, Cengage Learning. 2013. ISBN (10): 1-133-60448-X; ISBN (13): 978-1-133-60448-8

TEXTBOOK

Whitney, Eleanor Noss, and Sharon Rady Rolfes. *Understanding Nutrition.* 13th ed. Belmont, Calif.: Wadsworth, Cengage Learning. 2013. ISBN (10): 1-133-58752-6; ISBN (13): 978-1-133-58752-1

VIDEO PROGRAMS

Nutrition Pathways

Each video program is correlated with the student course guide and the reading assignment for that lesson. Be sure to read the Lesson Focus Points in the student course guide before you watch the program. Since examination questions will be taken from the video programs as well as from the reading, careful attention to both is vital to your success.

INTERACTIVE COURSE

Self-graded interactive exercises, pre- and post-self-assessments, and case-based, problem-solving scenarios are available to students whose institutions have opted to offer these. These activities are useful for reinforcement and review of lesson content and learning objectives. The interactive activities are offered in two formats: DVD-ROM and online. Ask your instructor how to access these activities if they are listed in your syllabus as a course requirement.

Course Guidelines

Follow these guidelines as you study the material presented in each lesson:

1. LESSON ASSIGNMENTS—
   Review the Lesson Assignments in order to schedule your time appropriately.

2. OVERVIEW—
   Read the Overview for an introduction to the lesson material.

3. LEARNING OBJECTIVES—
   Review the Learning Objectives and the lesson material that relates to them.

4. TEXT FOCUS POINTS—
   To get the most from your reading, review the Text Focus Points before reading the assignment. You may want to write responses or notes to reinforce what you have learned.

5. VIDEO FOCUS POINTS—
   To get the most from the video segment of the lesson, review the Video Focus Points before watching the video. You may want to write responses or notes to reinforce what you have learned.

6. PROJECT—
   The Diet Analysis Project is designed to provide you with specific information about nutrient adequacy and balance, number of kcalories consumed, nutrient density, and variety of the foods you typically eat. Consult the Table of Contents for the location of the Project forms.

7. GETTING IT TOGETHER—
   The purpose of the "Getting It Together" section is to provide an outline of *new* or *difficult* material presented in the current lesson. This section provides practice in recalling lesson material and utilizes the textbook interactively. Use this section as a learning aid, to reinforce knowledge, or to organize information introduced in the lesson.

8. OPTIONAL COMPUTER-BASED ACTIVITIES—
These computer-graded interactive exercises and activities are available for reinforcement and review of lesson content and learning objectives. Consult your instructor and/or syllabus for assigned activities.

9. RELATED ACTIVITIES—
The Related Activities are not required unless your instructor assigns them. They are offered as suggestions to help you learn more about the material presented in the lesson.

10. PRACTICE TEST—
Complete the Practice Test to help you check your understanding of the lesson.

11. ANSWER KEY—
Use the Answer Key at the end of the lesson to compare your answers and locate material related to each question of the Practice Test.

Lesson 1

Nutrition Basics

LESSON ASSIGNMENTS

Text: Whitney and Rolfes, *Understanding Nutrition*, Chapter 1, "An Overview of Nutrition," pp. 3–33; Chapter 2, "Planning a Healthy Diet," pp. 35–67 (Read the information from the first two chapters in an overview fashion *only*. Specific nutrients and topics will be covered in detail in subsequent chapters.)

Video: "Nutrition Basics" from the series *Nutrition Pathways*

Related Activities:

These activities are not required unless your instructor assigns them. They are offered as suggestions to help you learn more about the material presented in this lesson. Refer to your syllabus to determine which of these activities have been assigned.

Complete the attached forms for the following related activities, and return them to your instructor according to the established deadline.

- ❑ "Calculating the Energy Found in Foods"
- ❑ "Learning about the USDA MyPlate"
- ❑ "Learning about the U.S. Exchange Lists"
- ❑ "Applying the U.S. Exchange System"
- ❑ "Reading a Food Label to Calculate a Personal Daily Value for Fat"
- ❑ "Practice in Reading Food Labels"

Optional Web Activities:

Consult your instructor and/or syllabus for any assigned activities.

OVERVIEW

Nutrition plays a major role in our lives—from before we are born and into our golden years. This lesson demonstrates how the daily food choices we make throughout our lives influence how well we live and possibly how long we will be around. Choosing more nutrient-dense foods or less nutrient-dense, higher calorie foods affects our health—either positively or negatively. You will learn how the development of nutrition as a science affects what we eat and how food choices reflect family traditions, religious beliefs, emotions, financial status, socialization, food preferences, and nourishment. How do you go about making reasonable choices from the many foods available to you? Do you select foods based solely on nutritional value and disregard the other reasons why we choose foods? These questions and others are answered in Lesson 1.

This lesson introduces you to diet-planning principles and guidelines. These guidelines help you select foods that provide energy without excess calories and foods that provide variety, balance, and nutrient adequacy.

LEARNING OBJECTIVES

Upon completing this lesson, you should be able to:

1. Describe factors that affect food choices.

2. Describe the six major classes of nutrients.

3. Define the science of nutrition.

4. Describe how the development of nutrition as a science has influenced what people eat.

5. Describe the Dietary Reference Intakes.

6. Describe energy and nutrient recommendations.

7. Identify methods for assessing nutrient status.

8. Cite the ten leading causes of illness and death in the United States.

9. Explain diet-planning principles as they apply to food selection.

10. Cite the dietary guidelines for Americans.
11. Differentiate between food group plans.
12. Explain the function of the exchange lists.
13. Describe the components of a food label.

TEXT FOCUS POINTS

The following focus points are designed to help you get the most from your reading. Review them, then read the assignment. You may want to write notes to reinforce what you have learned.

1. What are the various factors that affect our food choices? What are functional foods? What are phytochemicals?

2. What are the six classes of nutrients? Define the terms *inorganic* and *organic*. What makes a nutrient "essential"? What are the energy-yielding nutrients? Which nutrients do not yield energy when broken down? How many kcalories are in each energy-yielding nutrient? Define *kcalorie* and *energy density*.

3. How do vitamins differ from the energy-yielding nutrients? How do minerals differ from the energy-yielding nutrients? What is the function of water in the body?

4. Define the *science of nutrition*. What is nutritional genomics? What four questions should people ask before concluding that an experiment has shown that a nutrient cures a disease or alleviates a symptom?

5. What do the Dietary Reference Intake values include? Define *Estimated Average Requirement*; *Recommended Dietary Allowance*; *Adequate Intake*; and *Tolerable Upper Intake Level*. How is the RDA for a nutrient set? Describe how the estimated average requirements (EAR) differ from the RDA. Define EER and the AMDR. Cite the five ways in which nutrient recommendations should be used.

6. What information is provided by the following nutrient assessment methods: historical information, anthropometric measurements, and laboratory tests? What is the goal of Healthy People?

7. What is the relation between risk factors and disease? Define *risk factor*. List the ten leading causes of death in the United States.

8. Define the following diet-planning principles of food selection: adequacy, balance, kcal control, nutrient density, moderation, and variety.

9. Cite the four major topic areas of the Dietary Guidelines for Americans.

10. What is the function of the USDA Food Patterns? Define *food group plans*. Define *discretionary kcalories*.

11. What is the function of the USDA's MyPlate? What do the four sections of the plate represent? How would an individual know their recommended number of servings from each food group? How does the recommendations from the USDA MyPlate and the USDA Food Patterns compare to actual intakes?

12. How is food sorted in exchange lists? Why were exchange lists originally developed?

13. What information should every food label display? In what order are ingredients given on a food label? List the "Nutrition Facts" that appear on labels. What are the percent Daily Values? What relationships for health claims have been approved by the FDA for use on food labels? What are structure-function claims?

VIDEO FOCUS POINTS

The following focus points are designed to help you get the most from the video segment of this lesson. Review them before watching the video. You may want to write notes to reinforce what you have learned.

1. What did nutrition scientists develop during the 1940s to help the nation improve nutrient status and prevent deficiencies? What is nutritional genomics? What was the purpose of the Food Guide Pyramid? What factors influence food choices?

2. What nutrient provides us with quick energy and is the main fuel source for the brain? What nutrient is considered the "building block" of life? What nutrient provides us with calories and carries essential vitamins to cells? What classes of nutrients help to activate chemical reactions in the body and assist many bodily functions? What nutrient is necessary to support every chemical reaction in the body and is needed in the greatest quantity?

3. What foods are represented on each level of the pyramid? How could a vegetarian use the Food Guide Pyramid? How much is in a serving of the following foods: pasta, raw vegetables, fruit, milk, and meat? How can guidelines for the USDA Food Guide help reduce the incidence of obesity?

4. Describe briefly the health concerns of each *Pathways* subject who will be followed for one year.

Calculating the Energy Found in Foods

NAME: ______________________________ ID #: ________________

The following exercises will help you learn how to calculate the energy available from food.

Refer to Chapter 1 in your textbook, "How to Calculate the Energy Available from Foods," for a sample calculation.
Remember: carbohydrate = 4 kcal/g; protein = 4 kcal/g; fat = 9 kcal/g.

PROBLEM #1:
If you ate a snack consisting of two peanut butter cookies and one-half cup of whole milk (19 grams of carbohydrate, 6 grams of protein, and 11 grams of fat), what would be the total number of kcalories you would consume?

1. ____________ kcal from carbohydrate *(Show your calculations below.)*

2. ____________ kcal from protein (*Show your calculations below.*)

3. ____________ kcal from fat (*Show your calculations below.*)

4. ____________ Total kcalories

—Continued on the back

PROBLEM #2:
What percentage of kcalories comes from each of the above nutrients?
(Divide the kcal from each nutrient by the total kcalories and then multiply by 100 to get the percentage. Round your answers to the nearest percent. Your total should equal about 100 percent.)

5. ____________ kcal from carbohydrate (*Show your calculations below.)*

6. ____________ kcal from protein (*Show your calculations below.*)

7. ____________ kcal from fat (*Show your calculations below.*)

8. The *Acceptable Macronutrient Distribution Ranges* (AMDR) state that the intake of fat should be between 20–35% of total kcalories for the day.

Learning about the USDA MyPlate

NAME: ______________________________ ID#: ________________________

The purpose of this activity is to help you obtain specific information about the *MyPlate* as a meal-planning tool.

Refer to the information on *Diet-Planning Guides* in Chapter 2 of your textbook to help you complete this assignment.

Cite page numbers in the text to support answers to the following questions:

1. Briefly explain the function of MyPlate.

2. Using a computer, go to www.choosemyplate.gov and access Super Tracker to create a profile. List the recommended range of servings for each food group below:

   Grains ______________________

   Vegetables ______________________

   Fruits ______________________

   Dairy ______________________

   Protein foods ______________________

   Oils ______________________

3. Choosing a sample meal plan from your Super Tracker profile, what foods would you eat to meet your targets?

—*Continued on the back*

4. Provide one example and its serving size for each food group below:

Grains ____________________

Vegetables ____________________

Fruits ____________________

Dairy ____________________

Protein foods ____________________

Oils ____________________

5. What do the colors represent on MyPlate?

6. What is the recommendation for physical activity for adults 18–64 years old?

Learning about the U.S. Exchange Lists

NAME: ______________________ ID#: ______________

The purpose of this activity is to help you obtain specific information about the exchange system.

Refer to the information on *Diet Planning Guides* in Chapter 2 and *Exchange Lists* in Appendix G in your textbook to help you complete this assignment.

Cite page numbers in the text to support answers to the following questions:

1. For what segment of the population was the exchange system originally developed?

2. In brief terms, describe how the exchange system sorts foods compared to the USDA Food Patterns.
   a. Exchange system:

   b. USDA Food Patterns:

3. Fill in the Table:
   a. Name the exchange lists found in the U.S. Exchange System.
   b. Provide one example of portion size for each list.
   c. Cite the number of grams of carbohydrate, protein, and fat associated with each portion size.
   d. Cite the total kcalories associated with each portion size.

U.S. EXCHANGE SYSTEM					
Exchange List	Portion Size	G Carb.	G Prot.	G Fat	Kcal

—Continued on the back

4. What is the primary difference in the exchange system with regard to meat/meat substitutes compared to the USDA Food Patterns?

   a. Exchange system:

   b. USDA Food Pattern:

5. Explain where each of the following foods are placed in the exchange lists and why they are placed there:

   a. Cheese:

   b. Corn, green peas, and potatoes:

   c. Olives, bacon, and avocados:

6. Briefly explain the benefits of combining the USDA Food Patterns and the exchange lists when designing an eating plan.

Applying the U.S. Exchange System

NAME: ______________________________ ID#: ______________________________

The purpose of this activity is to give you practice applying the exchange system as a meal-planning tool.

Fill in the Table: Assume the following meals represent a day of food for you. Refer to the *United States: Exchange Lists* in Appendix G of your textbook to estimate the number of servings and the total kcalories for the following meals:

BREAKFAST:	Starch Servings	Fruit Servings	Vegetable Servings	Milk Servings	Meat Servings	Fat Servings	TOTAL Calories
1 cup oatmeal							
1 cup skim milk							
1 small apple							
LUNCH:	Starch Servings	Fruit Servings	Vegetable Servings	Milk Servings	Meat Servings	Fat Servings	TOTAL Calories
2 oz. lean roast beef							
2 slices bread							
1 tbs. mustard							
20 fat-free tortilla chips							
DINNER:	Starch Servings	Fruit Servings	Vegetable Servings	Milk Servings	Meat Servings	Fat Servings	TOTAL Calories
4 oz. chicken breast without skin							
½ cup green beans							
1 cup brown rice							
1 cup low-fat ice cream							
TOTALS FOR THE DAY:							

—Continued on the back

Applying the U.S. Exchange System—*Continued*

1. Provide your personal opinion of the positive and negative aspects of the Exchange System in general.

   a. Positive aspects:

   b. Negative aspects:

2. Explain whether you would use the exchange system to design your personal eating plan.

Reading a Food Label to Calculate a Personal Daily Value for Fat

NAME: ______________________________ ID #: ____________________________

This exercise is designed for two purposes: to help you learn to read a food label for fat content and to help you calculate a personal Daily Value for fat.

Refer to the sample food label found in Chapter 2.

1. The % Daily Value column on a food label applies to what number of kcal/day?

2. Identify ten nutrients recorded as a % Daily Value on food labels.

____________________ ____________________ ____________________

____________________ ____________________ ____________________

____________________ ____________________

____________________ ____________________

3. ____________ What is your RDA for energy (kcal/day) according to the RDA table found in the textbook?

4. ____________ What is your daily allowance for fat kcalories? Multiply your RDA for energy by 0.30. (*Show your calculations below.*)

5. ____________ What is your daily allowance for fat grams? Divide total fat kcalories (step 4) by 9 kcal/g. (*Show your calculations below.*)

6. Assume *your* personal daily allowances (Steps 3–5 above) are different than the Daily Values found on a food label. What would account for those differences? Cite page numbers in your text to support your answer.

Practice in Reading Food Labels

NAME: ______________________________ ID #: ____________________________

This exercise is designed to give you practice reading food labels to obtain helpful information.

DIRECTIONS:

a. Select three different food product labels. Make sure at least two labels contain some fat.
b. Make copies of this form and record information from each label by answering the questions below.
c. Staple each label to the top of the copied form.
d. Staple all labels and completed questions together, and submit to your instructor.

1. What is the predominant ingredient?

2. What sweeteners are listed?

3. How many grams of total carbohydrate and dietary fiber are in the product?

4. Identify any fat or oil listed, and state whether it is predominantly saturated or unsaturated.

5. How many fat kcalories are in this product, and how does this compare to the recommendation that states less than 30 percent of kcal/day should come from fat? *(Show your calculations below.)*

6. How many milligrams of cholesterol are in the product, and what is the source of cholesterol?

—Continued on the back

7. How many grams of protein are in the product, and what is the source of protein?

8. What constitutes a serving size, and how many kcal are in one serving?

9. What, if any, stated or implied health claim is made about the product?

10. What did you learn about the product?

PRACTICE TEST

The following items will help you check your understanding of this lesson. Compare your answers to the Answer Key at the end of the lesson. Review the course materials related to any incorrect answer.

Multiple Choice: Select the one choice that best answers the question.

1. Which of the following represents a food choice based on negative association?
   A. A tourist from China who rejects a hamburger due to unfamiliarity
   B. A child who spits out his mashed potatoes because they taste too salty
   C. A teenager who grudgingly accepts an offer for an ice cream cone to avoid offending a close friend
   D. An elderly gentleman who refuses a peanut butter and jelly sandwich because he deems it a child's food

2. Which of the following nutrients does NOT yield energy during its metabolism?
   A. Fat
   B. Protein
   C. Vitamins
   D. Carbohydrates

3. Gram for gram, which of the following provides the most energy?
   A. Fats
   B. Alcohol
   C. Proteins
   D. Carbohydrates

4. What is the kcalorie value of a meal supplying 110 g of carbohydrates, 25 g of protein, 20 g of fat, and 5 g of alcohol?
   A. 160
   B. 345
   C. 580
   D. 755

5. Recommended Dietary Allowances may be used to __________
   A. measure nutrient balance.
   B. assess dietary nutrient adequacy.
   C. treat persons with diet-related illness.
   D. calculate exact food requirements for most individuals.

6. The recommendation for energy that is set at the average of the population's estimated requirement is __________
   A. acceptable macronutrient distribution range.
   B. adequate intake.
   C. estimated energy requirement.
   D. estimated average intake.

7. The Dietary Reference Intakes may be used to __________
   A. treat people with diet-related disorders.
   B. assess adequacy of all required nutrients.
   C. plan and evaluate diets for healthy people.
   D. assess adequacy of only vitamins and minerals.

8. In setting Dietary Reference Intakes for nutrients, the DRI Committee makes all of the following assumptions EXCEPT __________
   A. people generally are healthy.
   B. people generally consume protein of good quality.
   C. people generally consume diets adequate in kcalories.
   D. people buy their foods exclusively at health food stores.

9. Inspection of hair, eyes, skin, and posture is part of the nutrition assessment component known as __________
   A. diet history.
   B. anthropometrics.
   C. biochemical testing.
   D. physical examination.

10. What are the six diet-planning principles of diet planning?
   A. Abundance, B vitamins, kcalories, diet control, minerals, and variety
   B. Abundance, balance, conservative, diversity, moderation, and vitamins
   C. Adequacy, bone development, correction, vitamin density, master, and variety
   D. Adequacy, balance, kcalorie control, nutrient density, moderation, and variety

11. Nutrient density refers to foods that __________
    A. carry the USDA nutrition labeling.
    B. are higher in weight relative to volume.
    C. contain a mixture of carbohydrate, fat, and protein.
    D. provide more nutrients than kcalories relative to the RDA.

12. In planning a healthy diet, all of the following tools would be helpful EXCEPT __________.
    A. an internet website
    B. USDA Food Patterns
    C. choosemyplate.gov
    D. exchange system

13. Food exchange systems were originally developed for people with __________
    A. diabetes.
    B. terminal diseases.
    C. cardiovascular disease.
    D. life-threatening obesity.

14. According to nutrition labeling laws, the amounts of what two vitamins must be listed on the package label?
    A. Vitamins D and E
    B. Vitamins A and C
    C. Thiamin and riboflavin
    D. Vitamins B_6 and niacin

15. During the 1940s, to help the nation improve their nutrient status, nutrition scientists developed __________
    A. the Food Guide Pyramid.
    B. Diet-Planning Principles.
    C. Recommended Dietary Allowances.
    D. the concept of "vitamins."

16. Proteins are considered the__________
    A. main source of quick energy for the body.
    B. building blocks of life.
    C. activator of chemical reactions in the body.
    D. best source of concentrated energy.

17. Nearly every chemical reaction in the body is activated by __________
    A. fat.
    B. carbohydrate.
    C. vitamins.
    D. water.

18. What foods form the foundation for the USDA Food Guide?
    A. Fruits and vegetables
    B. Grains and cereals
    C. Dairy products
    D. Meat products

19. A serving of raw vegetables versus cooked vegetables is __________
    A. one cup raw versus one-half cup cooked.
    B. one-half cup raw versus one cup cooked
    C. equal.
    D. different depending on whether it is fresh or frozen.

Fill in the Blank: Insert the correct word or words in the blank for each item.

20. According to the USDA MyPlate, __________ constitutes a serving of milk.

21. The Pathway subjects illustrate how nutrition affects health concerns like __________, __________, and __________.

Matching: Select the letter next to the word or phrase that best matches the numbered statements. Answers are used only once.

______22. Substance containing no carbon, hydrogen, or oxygen

______23. Energy (kcal) required to increase temperature of one liter of water from 0°C to 100°C

______24. Excess nutrient intake leads to this

______25. A serving in the USDA Food Patterns

______26. Common cow's milk fortification nutrient

______27. Health claim allowed on food label

A. 85
B. 100
C. Iron
D. Sodium and hypertension
E. Protein
F. Vitamin A and D
G. 1 tbsp.peanut butter
H. Overnutrition
I. Anthropometrics

Essay: Reflect on the item(s) listed below; one or more may be included on the exam.

28. Explain how food choices are influenced by habits, emotions, physical appearance, and ethnic background.

29. A. List and discuss four methods commonly used to assess nutritional status of individuals.
    B. What types of individuals are qualified to evaluate nutritional health of individuals?

30. List the five food groups and describe how foods are classified. What are the advantages and disadvantages of the USDA Food Patterns and the MyPlate?

31. Describe the major aspects of nutrition labeling regulations. List the information that must be displayed on food labels.

ANSWER KEY

The following provides the answers and references for the Practice Test questions.

	Answer	Learning Objective	Reference
1.	D	LO 1	textbook, p. 5
2.	C	LO 2	textbook, p. 11
3.	A	LO 2	textbook, pp. 8–9
4.	D	LO 2	textbook, pp. 8–9
5.	B	LO 5	textbook, p. 18
6.	C	LO 6	textbook, p. 18
7.	C	LO 5	textbook, p. 17
8.	D	LO 5	textbook, p. 17
9.	D	LO 7	textbook, p. 23
10.	D	LO 9	textbook, p. 36
11.	D	LO 9	textbook, p. 36
12.	A	LO 11	textbook, pp. 38–44
13.	A	LO 12	textbook, p. 46
14.		LO 13	textbook, p. 55
15.	C	LO 4	Video
16.	B	LO 2	Video
17.	C	LO 2	Video
18.	B	LO 11	Video
19.	A	LO 11	Video
20.	1 cup milk	LO 11	Video
21.	Type 2 diabetes (diabetes), high blood cholesterol (CVD, heart disease), and weight loss	LO 8, 10, 11	Video
22.	C	LO 2	textbook, p. 7
23.	B	LO 2	textbook, p. 8
24.	H	LO 3	textbook, p. 21
25.	G	LO 7	textbook, p. 43
26.	F	LO 13	textbook, p. 49
27.	D	LO 13	textbook, p. 57
28.		LO 1	textbook, pp. 4–6
29.		LO 7	textbook, pp. 22–27
30.		LO 11	textbook, pp. 40–47
31.		LO 13	textbook, pp. 53–58

Lesson 2

The Digestive System

LESSON ASSIGNMENTS

Text: Whitney and Rolfes, *Understanding Nutrition*, Chapter 3, "Digestion, Absorption, and Transport," pp. 69–93

Video: "The Digestive System" from the series *Nutrition Pathways*

Getting It Together:
The purpose of the "Getting It Together" section is to provide an outline of *new* or *difficult* material presented in the current lesson. This section provides practice in recalling lesson material and utilizes the textbook interactively. Use this section as a learning aid, to reinforce knowledge, or to organize information introduced in the lesson.

Optional Web Activities:
Consult your instructor and/or syllabus for any assigned activities.

OVERVIEW

This lesson serves as an introduction to the digestive system. Subsequent lessons delve more deeply into digestion and absorption of the individual energy-yielding nutrients. This overview of digestion examines the common pathways that all foods take—from the time you put food in your mouth, to the transformation of foods into usable nutrients, to the transport of absorbed nutrients to the cells, to the final elimination of the unused nutrients and/or waste products. This lesson also includes problems encountered during digestion, and you will learn how the body solves those problems.

LEARNING OBJECTIVES

Upon completing this lesson, you should be able to:

1. Identify the basic route followed by food through the GI tract.

2. Describe the muscular action of digestion.

3. List the digestive organs and glands and their secretions that promote the breakdown of food.

4. Briefly describe the anatomy of the absorptive system.

5. Describe the basic transportation routes absorbed nutrients take in order to be delivered and used by the body.

6. Explain how the body regulates digestion and absorption.

7. Identify common digestive problems, explain how the GI tract is involved, and list common solutions to the problems.

TEXT FOCUS POINTS

The following focus points are designed to help you get the most from your reading. Review them, then read the assignment. You may want to write notes to reinforce what you have learned.

1. What is the GI tract? Where does the process of digestion begin? What is the pharynx? What is the function of the epiglottis? What is a bolus? What is the function of the esophagus? What is the function of the esophageal sphincter? What is chyme? What is the function of the pyloric sphincter? What is the function of the ileocecal valve? What is the function of the rectum?

2. What is peristalsis and where does it begin? What is segmentation? What is the function of a sphincter muscle?

3. What glands begin the process of digestion? What nutrient is partially digested by salivary enzymes? What is the action of hydrochloric acid? What is the function of mucus? What is the pH of the stomach? What does pancreatic juice contain? What organ produces bile? Where is bile stored? What is the

action of bile? What is the pH of the intestines? What is intestinal flora and what is its function? What is the function of fiber in digestion?

4. Where does the primary absorption of nutrients take place in the body? What are villi? What are microvilli and how do they function? What is the function of the crypts? What are goblet cells?

5. Which nutrients are transported directly through the vascular (bloodstream) system? Which nutrients are transported through the lymphatic system, and how do they enter the blood circulatory system? What are the primary differences between the two transport systems in the body?

6. Define the following terms: *hormone*, *gastrin*, *secretin*, *cholecystokinin*, and *gastric-inhibitory peptide*.

VIDEO FOCUS POINTS

The following focus points are designed to help you get the most from the video segment of this lesson. Review them before watching the video. You may want to write notes to reinforce what you have learned.

1. What are the signs of choking? What technique could you apply to assist a choking person?

2. What is the function of belching, and what foods may cause an increase in belching? What causes flatus, and what distinguishes it from bloating?

3. What is heartburn/acid indigestion? What could help alleviate the symptoms of heartburn/acid indigestion? What is the danger associated with vomiting?

4. What is constipation? How can constipation be prevented or treated?

5. What is the cause of peptic ulcers?

GETTING IT TOGETHER

Introduction to the Digestive System

Refer to Chapter 3 for a graphic representation of the GI tract and then fill in the blanks below to help you develop an understanding of the basic route of digestion through the GI tract.

THE GI TRACT:

1. The digestive process begins in the ____________.
2. The enzyme, _____________, partially digests starch.
3. The _____________ prevents food from entering the trachea.
4. A portion of chewed, swallowed food is known as a _____________.
5. The _____________conducts food past the diaphragm.
6. The sphincter muscle that keeps the bolus and acid in the stomach is the _____.
7. The _____________adds acidic juices to the bolus and grinds it.
8. The partially digested, semi liquid mass is known as _____________.
9. The sphincter muscle that keeps chyme in the small intestine is the __________.
10. Digestion and absorption of nutrients occurs primarily in the ____________.
11. The _____________stores and drips bile into the small intestine.
12. The _____________stores and drips digestive juices into the small intestine.
13. The sphincter muscle that keeps chyme in the large intestine is the ________.
14. The colon or bowel is also known as the _____________.
15. The muscle that holds waste in the large intestine is the _____________.
16. The last sphincter in the GI tract that allows voluntary passage of waste is the _____________.

Answer Key to Fill in the Blanks:

1. Mouth
2. Salivary amylase
3. Epiglottis
4. Bolus
5. Esophagus
6. Esophageal sphincter
7. Stomach
8. Chyme
9. Pyloric sphincter
10. Small intestine
11. Gallbladder
12. Pancreas
13. Ileocecal valve
14. Large intestine
15. Rectum
16. Anus

Refer to Chapter 3 to complete the outline below. Use the outline as a study aid and to reinforce material.

MUSCULAR ACTIONS:

- Peristalsis: ______________________________
- Segmentation: ______________________________
- Four major sphincter muscles: ______________________

DIGESTIVE SECRETIONS:

- Saliva: ______________________________
- Gastric juice: ______________________________
- Mucus: ______________________________
- Pancreatic juices: ______________________________
- Small intestine juices: ______________________________
- Bile: ______________________________

ABSORPTION:

- Small intestine: ______________________________
  - Villi ______________________________
  - Crypts ______________________________
  - Microvilli ______________________________
- Bloodstream: ______________________________
- Lymph: ______________________________

REGULATORY SECRETIONS:

- Homeostasis: ______________________________
- Hormones: ______________________________
  - Gastrin ______________________________
  - Secretin ______________________________
  - Cholecystokinin ______________________________
  - Gastric-inhibitory peptide ______________________________

PRACTICE TEST

The following items will help you check your understanding of this lesson. Compare your answers to the Answer Key at the end of the lesson. Review the course materials related to any incorrect answer.

Multiple Choice: Select the one choice that best answers the question.

1. What is one function of the pyloric sphincter?
   A. Secretes acid into the stomach
   B. Secretes hormones into the stomach
   C. Prevents the contents of the intestines from backing up into the stomach
   D. Prevents the contents of the intestine from emptying too quickly into the colon

2. What is a bolus?
   A. Enzyme that hydrolyzes starch
   B. Portion of food swallowed at one time
   C. Device used to analyze the contents of the stomach
   D. Sphincter muscle separating the stomach from the small intestine

3. Which of the following is a function of sphincter muscles?
   A. Control peristalsis
   B. Grind large food particles
   C. Secrete digestive juices into the GI tract
   D. Control the passage of food through the GI tract

4. What is a function of hydrochloric acid in the stomach?
   A. Absorbs water
   B. Inhibits peristalsis
   C. Neutralizes the food mass
   D. Creates an optimum acidity

5. What is one function of the gallbladder?
   A. Stores bile
   B. Produces bile
   C. Reabsorbs water and salts
   D. Performs enzymatic digestion

6. What is the name of the projections on the inner surface of the small intestine?
   A. Villi
   B. Cilia
   C. Mesenteric vessels
   D. Vascular projectiles

7. Which of the following is a function of the intestinal microvilli?
   A. Secretion of bile salts
   B. Secretion of digestive acid
   C. Transport of nutrient molecules
   D. Transport of pancreatic enzymes

8. Which of the following conducts lymph into the vascular system?
   A. Villi
   B. Mesentery
   C. Subclavian vein
   D. Common bile duct

9. Which of the following regulates the pH of the stomach?
   A. Gastrin
   B. Insulin
   C. Secretin
   D. Cholecystokinin

10. The use of an antacid is indicated PRIMARILY for which condition?
    A. Excessive gas
    B. Acid indigestion
    C. Excessive belching
    D. Bloating

11. Therapy for constipation would include all of the following EXCEPT __________.
    A. increasing water intake.
    B. decreasing fiber intake.
    C. increasing physical activity.
    D. responding promptly to the defecation signal.

Fill in the Blank: Insert the correct word or words in the blank for each item.

12. The most obvious sign of choking is that the person cannot ________________.

13. The function of belching is to rid the small intestine or stomach of ______________.

14. The greatest danger from vomiting is __________________.

Matching: Select the letter next to the word or phrase that best matches the numbered statements. Answers are used only once.

_____15. Prevents food from entering the windpipe when swallowing

_____16. Fingerlike projection of small intestinal lining

_____17. Type of cell that secretes mucus

_____18. Hormone produced by cells in the intestinal wall

A. Villus
B. Goblet cells
C. Epiglottis
D. Cholecystokinin
E. Insulin
F. Stomach

Essay: Reflect on the item(s) listed below; one or more may be included on the exam.

19. Name and describe the functions of the four major sphincter muscles that divide the GI tract into its principal regions.

20. Describe four common digestive problems and their recommended treatments or therapies.

ANSWER KEY

The following provides the answers and references for the Practice Test questions.

	Answer	Learning Objectives	References
1.	C	LO 1	textbook, p. 73
2.	B	LO 3	textbook, p. 72
3.	D	LO 2	textbook, p. 73
4.	D	LO 3	textbook, p. 74
5.	A	LO 1	textbook, pp. 70–71
6.	A	LO 4	textbook, p. 78
7.	C	LO 4	textbook, p. 78
8.	C	LO 5	textbook, p. 82
9.	A	LO 6	textbook, p. 83
10.	B	LO 7	textbook, p. 92
11.	B	LO 7	textbook, p. 91
12.	speak (make a sound, cough)	LO 7	Video
13.	carbon dioxide (gas)	LO 7	Video
14.	dehydration	LO 7	Video
15.	C	LO 1	textbook, p. 71
16.	A	LO 4	textbook, p. 78
17.	B	LO 3	textbook, p. 78
18.	F	LO 6	textbook, p. 84
19.		LO 2	textbook, pp. 72–74
20.		LO 7	textbook, pp. 88–93; video

Lesson 3

Carbohydrates: Simple and Complex

LESSON ASSIGNMENTS

Text: Whitney and Rolfes, *Understanding Nutrition*, Chapter 4, "The Carbohydrates: Sugars, Starches, and Fibers," pp. 95–126

Video: "Carbohydrates: Simple and Complex" from the series *Nutrition Pathways*

Getting It Together:

The purpose of the "Getting It Together" section is to provide an outline of *new* or *difficult* material presented in the current lesson. This section provides practice in recalling lesson material and utilizes the textbook interactively. Use this section as a learning aid, to reinforce knowledge, or to organize information introduced in the lesson.

Optional Web Activities:

Consult your instructor and/or syllabus for any assigned activities.

OVERVIEW

This lesson examines the carbohydrate family from its simplest form, sugar, to its most complex form, starch. The complex carbohydrate, fiber, is the topic of another lesson. You will learn how dietary carbohydrates are digested and absorbed, converted to glucose for immediate energy, converted to glycogen for reserve energy, and finally converted to fat for future energy stores. You will discover how blood glucose is regulated within the body and how insulin and glucagon work to maintain homeostasis. This lesson also examines the health effects of carbohydrates as well as the health implications of lactose intolerance and refined sugar intake.

LEARNING OBJECTIVES

Upon completing this lesson, you should be able to:

1. Identify the primary fuel source for the brain and nervous system and the foods that provide it.
2. Recognize the basic generic chemical formula of a simple carbohydrate.
3. Define simple carbohydrate and provide examples.
4. Define complex carbohydrate and provide examples.
5. Describe the basic steps involved in the digestion and absorption of starch.
6. Explain the health implications of lactose intolerance.
7. Describe the basic steps involved in the metabolism of absorbed starch.
8. Summarize the basic steps of blood glucose regulation.
9. Describe the health effects of sugar, and cite the recommended intake of sugar.
10. Describe the health effects of complex carbohydrate intake, and cite the recommended intake of simple and complex carbohydrates.

TEXT FOCUS POINTS

The following focus points are designed to help you get the most from your reading. Review them, then read the assignment. You may want to write notes to reinforce what you have learned.

1. What is the primary fuel for the brain? What is the storage form of glucose? What foods provide carbohydrates?
2. What is the basic chemical formula for a simple carbohydrate (monosaccharide)?
3. List the simple carbohydrates. Name the monosaccharides. What is the significance of glucose to nutrition? What monosaccharide is the sweetest? What foods contain fructose?

4. What does the term *disaccharide* mean? What is the common component of all disaccharides? What is the chemical reaction needed to make a disaccharide? What is the chemical reaction needed to take apart a disaccharide? Name the disaccharides. What is the most familiar disaccharide? What monosaccharide is found in milk sugar? What is the principal disaccharide found in milk sugar?

5. What does the term *polysaccharide* mean? What three complex carbohydrates are important in nutrition? What is the storage form of glucose in animals? What is the storage form of glucose in plants? Where is glycogen stored in the human body? What are the richest food sources of starch? (Reminder: Fiber will be the focus of another lesson.)

6. What is the ultimate goal of carbohydrate digestion and absorption? Where does starch digestion begin? What is the enzyme responsible for beginning starch digestion? What happens to starch digestion in the stomach? Where does most carbohydrate digestion occur? What major carbohydrate-digesting enzyme is responsible for further digesting polysaccharides to disaccharides and short glucose chains? Name the enzymes that dismantle the following disaccharides: maltose, sucrose, and lactose. What do all disaccharides contribute to the body after they are dismantled? After absorption occurs, what organ is responsible for converting fructose and galactose to glucose? What remains in the digestive tract after all sugars and most starches are digested? Where does most nutrient absorption take place?

7. What enzyme is absent or deficient in people suffering from lactose intolerance? What is the effect of aging on lactase activity? What percent of the population retain enough lactase to digest and absorb lactose efficiently? What are the symptoms and causes of lactose intolerance? What populations have the lowest and highest prevalence of lactose intolerance? How can lactose intolerance be managed?

8. What is the primary role of carbohydrate in human nutrition? What is the function of the liver in carbohydrate metabolism, and how does it respond to low blood glucose? How much of the body's total glycogen is stored in the liver and muscles?

9. Define *gluconeogenesis* and *protein-sparing action*. Under what circumstances are ketone bodies formed? How much dietary carbohydrate is necessary to spare body protein and prevent ketosis? What happens to glucose that is not used for immediate energy or converted to glycogen?

10. What is homeostasis? What can happen if blood glucose fluctuates to the extremes—either high or low? What are the two main regulatory hormones that control blood glucose concentration? Summarize the steps involved in blood glucose regulation (refer to Figure 4-12). How does epinephrine function in blood glucose regulation? What are two conditions that can result when blood glucose regulation fails? Define *hypoglycemia*. What is the "glycemic response" of food? Briefly explain the controversy surrounding the usefulness of the glycemic index.

11. How many teaspoons of sugar does the average person consume per day? What is the recommended sugar intake based on a percentage of total kcal per day? Briefly describe two ways in which excessive sugar intake can be detrimental to people. Explain health concerns from excessive sugar intake.

12. Describe the health effects of complex carbohydrates on the following: weight control, heart disease, cancer, and diabetes. What is the recommended intake of carbohydrate as a percentage of total energy? What is the recommended grams of carbohydrates established by the FDA on food labels?

VIDEO FOCUS POINTS

The following focus points are designed to help you get the most from the video segment of this lesson. Review them before watching the video. You may want to write notes to reinforce what you have learned.

1. What food source should provide most of the energy for active people? What are the long-term and immediate benefits of high carbohydrate diets to athletes?

2. How do carbohydrates affect blood glucose levels?

3. What is the link between sugar intake and hyperactivity and growth?

GETTING IT TOGETHER

Carbohydrates: Simple and Complex

Refer to Chapter 4 for graphic representations of simple and complex carbohydrates to complete the outline below, to serve as a study aid, and to reinforce material.

THE BASICS:

- Simple carbohydrates:
    - o Monosaccharides: ______________________________
    - o Disaccharides: ______________________________
- Complex carbohydrates:
    - o Polysaccharides: ______________________________
    - o Glycogen: ______________________________
    - o Starch: ______________________________
    - o Fibers: ______________________________
- Basic chemical formula for all simple carbohydrates: ____________

DIGESTION: (Review Chapter 3 in textbook—"Digestion, Absorption, and Transport")

Refer to Chapter 4 for a graphic representation of carbohydrate (CHO) digestion, then fill in the blanks below to follow CHO through the GI tract, through absorption, and through metabolism.

1. Starch digestion begins in the ____________ with a salivary enzyme known as ____________.

2. Hydrolysis of starch in the stomach is stopped due to the action of ____________ and ____________.

3. Starch digestion continues in the ____________ after it leaves the stomach.

4. Maltase hydrolyzes the disaccharide known as ____________.

5. Sucrase hydrolyzes the disaccharide known as ______________.

6. Lactase hydrolyzes the disaccharide known as ______________.

7. The large intestine contains indigestible carbohydrates known as ______________ and ______________ that are not digested by human enzymes, but that are digested by ______________.

8. Absorption occurs primarily in the ______________.

9. Some glucose absorption occurs in the ______________.

10. Absorbed monosaccharides enter the blood, go to the liver, and are converted to ______________.

11. The main energy source for the brain, nerve cells, and blood cells is ______________.

12. ______________ (fractional amount) of the body's total glycogen is stored in the liver.

13. ______________ (fractional amount) of the body's total glycogen is stored in the muscles.

14. The organ that dismantles glycogen to glucose for export into the blood is the ______.

15. The making of glucose from noncarbohydrate sources such as body protein is known as ______________.

16. ______________ spares protein from gluconeogenesis and prevents ketosis.

17. We need at least ______________g CHO per day to prevent ketosis and excessive breakdown of body protein.

18. Some glucose not used for immediate energy is stored as ______________.

19. Excess glucose not stored as glycogen is converted to ______________.

Answer Key to Fill in the Blanks:

1. Mouth; salivary amylase
2. Hydrochloric acid; protein-digesting enzymes
3. Small intestine
4. Maltose
5. Sucrose
6. Lactose
7. Fibers; resistant starch; bacterial enzymes
8. Small intestine
9. Mouth
10. Glucose
11. Glucose
12. One-third
13. Two-thirds
14. Liver
15. Gluconeogenesis
16. Carbohydrate
17. 50–100
18. Glycogen
19. Body fat (triglycerides)

Refer to Chapter 4 to complete the outline below. Use the outline as a study aid and to reinforce material.

REGULATING BLOOD GLUCOSE:

- Insulin: ______________________________
- Glucagon: ______________________________
- Epinephrine: ______________________________

HEALTH EFFECTS OF CARBOHYDRATE INTAKE:

- Excessive sugar: ______________________________
- Complex carbohydrates: ______________________________
- Recommended percentage of total kcal/day from carbohydrates:__________
- Recommended percentage of total kcal/day from refined sugar:__________
- Lactose intolerance: ______________________________

PRACTICE TEST

The following items will help you check your understanding of this lesson. Compare your answers to the Answer Key at the end of the lesson. Review the course materials related to any incorrect answer.

Multiple Choice: Select the one choice that best answers the question.

1. In which of the following are ample amounts of carbohydrates almost always found?
   A. Plant foods
   B. Health foods
   C. Animal products
   D. Protein-rich foods

2. Which of the following is NOT a simple carbohydrate?
   A. Starches
   B. White sugar
   C. Disaccharides
   D. Monosaccharides

3. Which of the following is known as blood sugar or dextrose?
   A. Glucose
   B. Maltose
   C. Sucrose
   D. Fructose

4. What is the reaction that links two monosaccharides together?
   A. Hydrolysis
   B. Absorption
   C. Disaccharide
   D. Condensation

5. What is the principal carbohydrate of milk?
   A. Lactose
   B. Sucrose
   C. Maltose
   D. Glycogen

6. Glycogen is stored mainly in which of the following tissues?
   A. Muscle and liver
   B. Pancreas and kidney
   C. Stomach and intestine
   D. Brain and red blood cells

7. What is the primary organ that converts fructose to glucose following absorption?
   A. Liver
   B. Pancreas
   C. Skeletal muscle
   D. Small intestines

8. Gluconeogenesis is a term that describes the synthesis of _________
   A. amino acids from glucose.
   B. lactose from a source of sucrose.
   C. fat from excess carbohydrate intake.
   D. glucose from a noncarbohydrate substance.

9. Which of the following is a feature of diabetes?
   A. Type 1 diabetes is also known as adult-onset diabetes.
   B. It is believed to be caused by abnormal intake of dietary carbohydrates.
   C. Insulin-dependent type is more common than the noninsulin-dependent type.
   D. Dietary management should focus on total carbohydrate intake rather than type of carbohydrate.

10. Athletes or active individuals who require large amounts of energy should consume a diet high in __________
   A. protein.
   B. fat.
   C. carbohydrate.
   D. water.

11. Which of the following statements is the most accurate regarding studies documenting hyperactivity and sugar intake in children?
   A. There is very little evidence to support the statement that sugar causes hyperactivity.
   B. There is considerable evidence that sugar and hyperactivity are positively linked.
   C. There is scientific proof that sugar causes hyperactivity.
   D. None of the above.

Fill in the Blank: Insert the correct word or words in the blank for each item.

12. Hypoglycemia refers to a low level of ________________.

Matching: Select the letter next to the word or phrase that best matches the numbered statements. Answers are used only once.

_____13. A complex carbohydrate in muscle

_____14. Site where digestion of disaccharides takes place

_____15. Substance that signals the release of glucose *into* blood

_____16. Normal blood glucose level, in mg per 100 ml blood

A. 100
B. 150
C. Small intestine
D. Glucagon
E. Glycogen

Essay: Reflect on the item(s) listed below; one or more may be included on the exam.

17. Describe the body's mechanisms for controlling blood glucose levels under normal and stress conditions.

18. List two concerns of dietary sugar.

ANSWER KEY

The following provides the answers and references for the Practice Test questions.

	Answer	Learning Objectives	References
1.	A	LO 4	textbook, p. 95
2.	A	LO 4	textbook, p. 96
3.	A	LO 1	textbook, p. 96
4.	D	LO 2	textbook, p. 98
5.	A	LO 3	textbook, p. 99
6.	A	LO 4	textbook, p. 99
7.	A	LO 5	textbook, p. 103
8.	D	LO 7	textbook, p. 105
9.	D	LO 8, 10	textbook, pp. 107–108
10.	C	LO 6	Video
11.	A	LO 6	Video
12.	blood sugar	LO 8	Video
13.	E	LO 4	textbook, p. 99
14.	C	LO 5	textbook, p. 101
15.	D	LO 8	textbook, p. 106
16.	A	LO 8	textbook, p. 106
17.		LO 8	textbook, pp. 106–108
18.		LO 9	textbook, pp. 110–112

Lesson 4

Carbohydrates: Fiber

LESSON ASSIGNMENTS

Text: Whitney and Rolfes, *Understanding Nutrition*, Chapter 4, "The Carbohydrates: Sugars, Starches, and Fibers," pp. 100–102 and pp. 115–121

Video: "Carbohydrates: Fiber" from the series *Nutrition Pathways*

Project: "Diet Analysis Project"

Related Activity:

This activity is not required unless your instructor assigns it. It is offered as a suggestion to help you learn more about the material presented in this lesson. Refer to your syllabus to determine whether this activity has been assigned.

Complete the enclosed form for the following related activity, and return it to your instructor according to the established deadline.

❑ "Learning about Fiber in Foods"

Optional Web Activities:

Consult your instructor and/or syllabus for any assigned activities.

OVERVIEW

"You need to eat your roughage," Grandma used to tell us as she put a plate of vegetables in front of us. When we look back on it, we realize that Grandma was very smart! The roughage she was referring to is the indigestible complex carbohydrate known as fiber found in vegetables, fruits, grains, and legumes.

This lesson not only examines the importance of fiber to the health of the body, but also explains what happens to dietary fiber during digestion and absorption and the associated adverse side effects of too much fiber in the diet.

Finally, this lesson explores the impact of artificial sweeteners in the diet and the different types of natural sweeteners used in the diet and their effects on health.

The end of this lesson marks the beginning of the Diet Analysis Project in which you will begin the actual process of analyzing your personal eating habits.

LEARNING OBJECTIVES

Upon completing this lesson, you should be able to:

1. Define *dietary fiber*, including the characteristics of the different types of fiber, and explain how fiber differs from starch and how fibers are classified.
2. Explain what phytic acid is and how it may impact mineral absorption.
3. Explain how fiber is digested and absorbed.
4. Describe the possible positive health effects of fiber in the diet.
5. Describe the possible adverse effects of fiber in the diet.
6. Cite the recommended intake of fiber, and explain how you can obtain adequate fiber by following the Daily Value set by the FDA.
7. Describe the pros and cons of artificial sweeteners.
8. Explain what happens to grain when it is refined into flour.

TEXT FOCUS POINTS

The following focus points are designed to help you get the most from your reading. Review them, then read the assignment. You may want to write notes to reinforce what you have learned.

1. What is dietary fiber? What are the major nonstarch polysaccharides? What is the main difference between starch and fiber with regard to their bonds? Define the following terms: *soluble fibers*, *viscous*, *fermentable*, *insoluble fibers*, *resistant starches*.
2. How do soluble fibers affect the body? How do insoluble fibers affect the body?

3. What is phytic acid? What might be the impact of a high-fiber diet on minerals in the body?

4. What is the effect of fiber in the stomach during digestion? What is the function of fiber in the large intestine? What is the effect of bacteria on fiber? What is the energy contribution of fiber?

5. How can fiber impact weight control? How do foods rich in soluble fiber lower blood cholesterol, thus reducing the risk for heart disease? Explain how a high-fiber diet may protect against colon cancer. What effect do soluble fibers have on diabetes with regard to glucose absorption? Explain how dietary fibers enhance the health of the large intestine with regard to the following: stool size and passage, transit time, constipation, water intake, hemorrhoids, and diverticula.

6. How can too much fiber affect a person with a small stomach capacity? What symptoms are associated with increasing fiber too quickly? What steps can be taken to prevent complications associated with increased fiber intake?

7. According to the FDA, what is the recommended intake of fiber per day? What is the average daily fiber intake for people in the United States? What foods and what number of servings from the USDA Food Patterns will ensure an adequate supply of fiber each day? List three health benefits from soluble fiber.

8. Define artificial sweeteners? What are sugar alcohols? What is the ADI? Which artificial sweeteners are approved for use in the United States?

VIDEO FOCUS POINTS

The following focus points are designed to help you get the most from the video segment of this lesson. Review them before watching the video. You may want to write notes to reinforce what you have learned.

1. When grain is refined, as in white flour, what part(s) of the grain is(are) removed? Compare grams of fiber in refined and unrefined grains.

2. What are the "famous five" conditions that fiber protects us against? How does fiber impact blood cholesterol, risk for heart attack, weight loss, and diabetes?

3. Explain fiber's effects on cancer.

4. How might high dietary fiber intakes impact people with Crohn's disease or other inflammatory bowel diseases?

5. What must people do to prevent constipation associated with large fiber intakes? What can happen to people who consume too much fiber or who add fiber too quickly to their diets?

6. What dietary recommendations are encouraged for people with type 2 diabetes regarding fiber and carbohydrate intakes?

PROJECT

This project is required in order to complete the course successfully.

This portion of the lesson is designed to provide you with specific information about nutrient adequacy and balance, number of kcalories consumed, nutrient density, and variety of the foods you typically eat. After analyzing a three-to-seven-day food intake, you will be able to apply the information you have learned to make appropriate changes to improve eating habits.

You will record all of the food you have eaten over a three-to-seven-day period. Record the food on the Daily Food Record (DFR) form provided in the Diet Analysis Project information at the back of this student course guide. You will then input data in a computer using a diet analysis software. If you do not have access to a computer and diet analysis program, you will analyze all data manually. You will continue to analyze data through Lesson 15. The completed project will be submitted by the end of the semester as determined by your course instructor.

Refer to the Diet Analysis Project for complete details and directions. Contact your instructor if you have any questions. The Diet Analysis Project is in the back of this student course guide.

The information provided in Chapter 4 will be useful when answering questions pertaining to your Diet Analysis Project with regard to carbohydrate and fiber intake.

Learning about Fiber in Foods

NAME: ________________________________ ID #: ________________________

This exercise is designed to give you practice calculating how much fiber is consumed in a given day.

Use the diet analysis software or refer to Appendix H in your textbook to complete the following activity. Refer to Chapter 4 to answer questions.

DIRECTIONS: Calculate how many grams of fiber are found in each of the following foods, then answer the questions that follow.

BREAKFAST	**GRAMS FIBER**
1 cup oatmeal	
1 cup skim milk	
1 small apple	
1 tsp sugar	
LUNCH	**GRAMS FIBER**
2 oz lean roast beef	
2 slices white bread	
1 tbs mustard	
1 oz potato chips	
DINNER	**GRAMS FIBER**
4 oz chicken breast	
½ cup green beans	
1 cup brown rice	
1 cup LF frozen yogurt	
TOTAL FIBER	

—Continued on the back

Learning about Fiber in Foods—*Continued*

Cite page numbers in the textbook to support answers to the following questions:

1. According to the FDA daily value, what is the recommended daily intake of fiber? What is the DRI recommendation for fiber?

2. Identify the *type* of fiber that will decrease the risk for heart disease and certain cancers.

   Heart disease: ____________________ Cancer: ____________________

3. Compare the total fiber found in the foods listed on the previous page to the recommendation stated in Step 1. Was the total fiber content within, below, or above the recommendation?

4. If the total fiber (Step 3) was below the recommendation, name three additional foods AND the amount of fiber in each food that would increase the total to the minimum fiber recommendation.

   a. ____________________

   b. ____________________

   c. ____________________

5. Cite two precautions or warnings to consider when increasing daily fiber intake.

PRACTICE TEST

The following items will help you check your understanding of this lesson. Compare your answers to the Answer Key at the end of the lesson. Review the course materials related to any incorrect answer.

Multiple Choice: Select the one choice that best answers the question.

1. What are cellulose, pectin, hemicellulose, and lignin?
   A. Fibers
   B. Starches
   C. Sugar alcohols
   D. Artificial sweeteners

2. Which of the following is an example of the difference between the chemical bonds in starch and those in cellulose?
   A. Starch bonds are single.
   B. Starch bonds are fatty acids.
   C. Cellulose bonds release energy.
   D. Cellulose bonds are not hydrolyzed by human enzymes.

3. Which of the following plays a major role in the breakdown of certain types of dietary fiber reaching the large intestines?
   A. Bacteria
   B. Pancreas
   C. Colonic cells
   D. Small intestinal villus cells

4. Water-soluble fibers include all of the following EXCEPT ________
   A. gums.
   B. pectins.
   C. lignins.
   D. psyllium.

5. Which of the following fibers is water-insoluble?
   A. Gums
   B. Pectins
   C. Cellulose
   D. Psyllium

6. According to the DRI, what is the recommended daily intake of dietary fiber?
   A. 25–35 g
   B. 40–50 g
   C. 55–70 g
   D. 75–100 g

7. Which of the following provides the most fiber?
   A. One orange
   B. One-half cup of oatmeal
   C. One-half cup of split peas
   D. One-half cup of green peas

8. Which of the following is a characteristic of dietary fiber?
   A. Causes diverticulosis
   B. Usually found in high-fat foods
   C. Raises blood cholesterol levels
   D. Classified according to solubility in water

9. Which of the following is a feature of the pectins?
   A. They are used to thicken jelly.
   B. They are classified as insoluble fibers.
   C. They are resistant to intestinal bacterial fermentation.
   D. They are found in the small seeds of fruits such as strawberries.

10. When we eat white bread, the part of the wheat grain that we are eating is the
    A. bran.
    B. germ.
    C. endosperm.
    D. all of the above.

11. Which of the following describes the compound phytic acid?
    A. Product of starch digestion
    B. Nonnutrient component of plant seeds
    C. Found in gastric juice and helps to lower pH of chyme
    D. Found in high concentrations in the blood of people with diabetes

12. High levels of dietary fiber intake probably protect us from _________
    A. colon cancer.
    B. rectal cancer.
    C. breast cancer.
    D. all of the above.

13. In patients with inflammatory bowel disease or Crohn's disease, high fiber intakes may benefit them by _________
    A. normalizing the intestinal contents throughout the colon.
    B. causing them to have lower blood pressure which accompanies Crohn's disease.
    C. preventing weight gain that accompanies inflammatory bowel disease.
    D. all of the above.

14. In order to increase fiber intake, people with type 2 diabetes would be encouraged to eat all of the following foods EXCEPT _________
    A. orange juice.
    B. whole oranges.
    C. potatoes.
    D. broccoli.

Fill in the Blank: Insert the correct word or words in the blank for each item.

15. The amount of fiber per serving found in whole wheat kernels and in refined wheat flour, respectively, is _________ and _________.

16. Three of the "famous five" conditions that fiber is known to protect against are ____.

17. When people increase their dietary fiber intake, they must simultaneously always increase their intake of _______________.

Matching: Select the letter next to the word or phrase that best matches the numbered statements. Answers are used only once.

______18. Safe sweetener except for people with PKU

______19. A water-insoluble fiber

______20. Site where fibers may be metabolized to short-chain fatty acids

A. Pectin

B. Lignin

C. Gum

D. Large intestine

E. Aspartame

Essay: Reflect on the item(s) listed below; one or more may be included on the exam.

21. Give several examples of soluble and insoluble dietary fibers. List food sources of these fibers. Contrast the physical characteristics and features of these two types of fiber and their effects on GI tract function.

22. List and discuss five benefits of fiber.

ANSWER KEY

The following provides the answers and references for the Practice Test questions.

	Answer	Learning Objectives	References
1.	A	LO 1	textbook, p. 100
2.	D	LO 1	textbook, p. 100
3.	A	LO 3	textbook, pp. 101–102
4.	C	LO 1	textbook, p. 117
5.	C	LO 1	textbook, p. 117
6.	A	LO 6	textbook, p. 118
7.	C	LO 6	textbook, p. 119
8.	D	LO 1	textbook, p. 100
9.	A	LO 1	textbook, p. 100
10.	C	LO 8	Video
11.	B	LO 2	textbook, p. 100
12.	D	LO 4	Video
13.	A	LO 4	Video
14.	A	LO 4	Video
15.	9 grams and 1 gram	LO 8	Video
16.	heart disease cancer obesity diabetes high blood pressure	LO 4	Video
17.	water	LO 5	Video
18.	E	LO 7	textbook, p. 114
19.	B	LO 1	textbook, p. 117
20.	D	LO 3	textbook, p. 102
21.		LO 1	textbook, pp. 100–102
22.		LO 4	textbook, p. 117

Lesson 5

Fats: The Lipid Family

LESSON ASSIGNMENTS

Text: Whitney and Rolfes, *Understanding Nutrition*, Chapter 5, "The Lipids: Triglycerides, Phospholipids, and Sterols," pp. 129–147

Video: "Fats: The Lipid Family" from the series *Nutrition Pathways*

Getting It Together:

The purpose of the "Getting It Together" section is to provide an outline of *new* or *difficult* material presented in the current lesson. This section provides practice in recalling lesson material and utilizes the textbook interactively. Use this section as a learning aid, to reinforce knowledge, or to organize information introduced in the lesson.

Optional Web Activities:

Consult your instructor and/or syllabus for any assigned activities.

OVERVIEW

When people think of fat, they usually think "I've got to reduce the fat in my diet!" or "Fat's bad for you—it causes heart disease!" We typically think of the negative health effects of too much fat in the diet. Fat, however, is not inherently bad. The fact is, without fat there would be very little of the flavor or aroma that we associate with fatty foods. As you explore digestion, absorption, and transportation of fat, you will examine the lipoproteins and their functions. You will learn about fatty acid saturation and how it affects stability and the shelf life of foods. You will also identify sources of essential fatty acids, including omega-3 fatty acids. Lipid metabolism is previewed in this lesson, with more detail given in Lesson 9. Finally, you will examine the origin and function of fat substitutes and how they are used in the diet.

LEARNING OBJECTIVES

Upon completing this lesson, you should be able to:

1. Identify the members of the lipid family.
2. Explain the basic function of lipids in the body and in foods, including which lipid predominates in the body and in foods.
3. Recognize the basic chemical structure of each member of the lipid family.
4. Explain the differences between saturated, monounsaturated, and polyunsaturated fatty acids, providing examples of food sources for each.
5. Identify the essential fatty acids, including sources of omega-3 and omega-6 fatty acids.
6. Describe the process and controversy surrounding hydrogenation.
7. Describe the basic steps involved in the digestion, absorption, and transport of lipids.
8. Explain the composition, function, and fate of the lipoproteins—chylomicrons, very-low-density lipoproteins (VLDL), low-density lipoproteins (LDL), and high-density lipoproteins (HDL).
9. Briefly explain how fat is metabolized.
10. Explain the impact of cholesterol and triglycerides on health.
11. Describe the origin and function of fat substitutes.

TEXT FOCUS POINTS

The following focus points are designed to help you get the most from your reading. Review them, then read the assignment. You may want to write notes to reinforce what you have learned.

1. What are the members of the lipid family? Which lipid predominates in foods and in the body?

2. Define *fatty acid*. Upon what does the degree of saturation depend? Define the following fatty acids: *saturated*, *monounsaturated*, and *polyunsaturated*. What is the predominant fatty acid found in olive oil? What is the predominant fatty acid found in corn oil?

3. What is the chemical composition of a triglyceride?

4. What chemical property of fatty acid influences the firmness of fats at room temperature? Generally speaking, which fatty acids are more liquid at room temperature, and what is the source of these fatty acids? Which fatty acids are harder at room temperature, and what is the source of these fatty acids? Name two vegetable oils that are highly saturated.

5. What happens to fat when exposed to oxygen? Which fatty acids spoil most easily due to oxidation? What is hydrogenation? What are two advantages and one disadvantage of hydrogenation? What is a *trans*-fatty acid?

6. What compound is the best known phospholipid? What is the basic chemical structure of lecithin? Why are phospholipids useful in the food industry? What are the richest food sources of lecithin? Why is lecithin not considered an "essential" nutrient? How many kcalories per gram does lecithin contribute to energy intake?

7. What is the most famous sterol? What is the structure of a sterol? What foods contain sterols? What foods contain cholesterol? What other compounds in the body are made with cholesterol as the starting material? Why is cholesterol not considered an "essential" nutrient? What organ manufactures cholesterol?

8. What enzymes in the mouth play a small role in the digestion of triglyceride? Where are triglycerides digested? What is the function of bile? Where is bile stored in the body? What hormone signals the gallbladder to release bile into the GI tract? What is emulsification? What are the two destinations of bile?

9. What are micelles? What is the group of compounds used as transport vehicles for absorbed fats? Describe the basic composition and function of the following lipoproteins: chylomicron, VLDL, LDL, and HDL.

10. What are the functions of triglycerides in the body? What are the functions of triglycerides in foods?

11. What are the two essential fatty acids that the body cannot make? What are eicosanoids? What is the primary member of the omega-6 fatty acid family? What

other omega-6 fatty acid can be made from linoleic acid? What foods provide omega-6 fatty acids in the diet? What is the primary member of the omega-3 fatty acid family? What two other omega-3 fatty acids can be made from linolenic acid? What are good food sources of omega-3 fatty acids? What are good sources of EPA (eicosapentaenoic acid) and DHA (docosahexaenoic acid) in the diet? Briefly explain the effects of fatty acid deficiencies.

12. What is lipoprotein lipase? Briefly describe how body fat is made from dietary carbohydrate, protein, and fat. How much of the body's ongoing energy needs during rest are provided by fat? What is hormone-sensitive lipase? What part of fat (triglyceride) can be used to provide glucose?

VIDEO FOCUS POINTS

The following focus points are designed to help you get the most from the video segment of this lesson. Review them before watching the video. You may want to write notes to reinforce what you have learned.

1. Why are triglycerides considered the best energy source for strenuous activities, such as hiking? What are examples of high-fat foods that would support an activity, such as hiking?

2. Why are lipids transported in lipoproteins?

3. When does cholesterol become harmful to the body? In order to deal with problems associated with cholesterol, what two components must be considered? What dietary component can significantly increase LDL, or the "bad" cholesterol? What lifestyle changes can improve blood cholesterol?

4. Why were fat substitutes developed? Define *Simplesse* and *olestra*.

5. What is the recommended limit of blood cholesterol? What dietary factors can increase triglyceride levels? What happens to an artery that has cholesterol deposits? What dietary components must be monitored for someone who has high blood cholesterol?

GETTING IT TOGETHER

Fats: The Lipid Family

Refer to Chapter 5 for graphic representations of a triglyceride, a phospholipid, and a sterol and complete the outline below. Use the outline as a study aid and to reinforce material.

FATTY ACIDS:

FATTY ACID SATURATION:

BASIC CHEMICAL FORMULA FOR TRIGLYCERIDES (FATS):

INFLUENCES OF DEGREE OF UNSATURATION:

HYDROGENATION:

- *Trans*-fatty acids:

BASIC CHEMICAL FORMULA FOR PHOSPHOLIPIDS:

ROLES OF PHOSPHOLIPIDS:

- Lecithin:

BASIC CHEMICAL FORMULA FOR STEROLS:

ROLES OF STEROLS:

- Cholesterol:

ROLES OF TRIGLYCERIDES:

ESSENTIAL FATTY ACIDS:

- Linoleic acid:
    - o Arachidonic acid
- Linolenic acid:
    - o EPA/DHA

EICOSANOIDS:

DIGESTION OF FATS: (Review Chapter 3— "Digestion, Absorption, and Transport")

Refer to Chapter 5 for a graphic representation of lipid digestion and then fill in the blanks below to follow lipids through the GI tract, through absorption, and through metabolism.

1. Some lipid digestion begins in the _______________ by an enzyme known as _______________.

2. Fat in the small intestine triggers the release of the hormone _______________ which signals the gallbladder to release _______________.

3. Complete digestion of lipids occurs in the _______________.

4. Fats must be _____________ by bile from the _____________ prior to digestion.

5. Spherical complexes known as _______________ form to carry digested fats into small-intestinal cells.

6. Fats are absorbed in the _______________ of the small-intestinal cells.

7. Fats are transported by _______________ to the rest of the body.

8. Lipoproteins known as _______________ are produced in the small intestine to transport *dietary* fat to cells.

---

9. The two lipoproteins produced in the liver are _______________ and _______________.

10. The lipoprotein that is produced from VLDL is _____________.

11. Lipoproteins consist of _______________, _______________, _______________, and _______________.

12. The least dense lipoprotein is a _____________.

13. The lipoprotein that is predominantly cholesterol is _____________.

14. The lipoprotein that is predominantly protein and the most dense is _____________.

15. The enzyme on cells that attracts triglycerides into cells is known as _____________.

16. _______________ is the enzyme in cells that hydrolyzes triglycerides, releasing fatty acids and glycerol into the blood.

Answer Key to Fill in the Blanks

1. Mouth; lingual lipase
2. CCK; bile
3. Small intestine
4. Emulsified; gallbladder
5. Micelles
6. Microvilli
7. Lipoproteins
8. Chylomicrons
9. VLDL; HDL
10. LDL
11. Protein, triglycerides, phospholipids, sterols
12. Chylomicron
13. LDL
14. HDL
15. Lipoprotein lipase
16. Hormone-sensitive lipase

PRACTICE TEST

The following items will help you check your understanding of this lesson. Compare your answers to the Answer Key at the end of the lesson. Review the course materials related to any incorrect answer.

Multiple Choice: Select the one choice that best answers the question.

1. What is the chemical composition of fats?
   A. Hexose polymers
   B. Glycogen granules
   C. Fatty acids and glycerol
   D. Combinations of long-chain fatty acids

2. Which type of fatty acid is found in high amounts in olive oil?
   A. Saturated
   B. Monounsaturated
   C. Polyunsaturated
   D. Partially hydrogenated

3. Characteristics of hydrogenated oils include all of the following EXCEPT ______
   A. they are stored in adipose tissue.
   B. they lower HDL and raise LDL cholesterol in the body.
   C. some of their fatty acids change shape from *cis* to *trans*.
   D. products containing them become rancid sooner, contributing to a shorter shelf life.

4. An oil that is partially hydrogenated sometimes changes one or more of its double-bond configurations from __________
   A. *cis* to *trans*.
   B. solid to liquid.
   C. covalent to ionic.
   D. saturated to unsaturated.

5. Which of the following is an omega-3 fat?
   A. Acetic acid
   B. Palmitic acid
   C. Linoleic acid
   D. Docosahexaenoic acid

6. Which of the following is a good source of eicosapentaenoic acid?
   A. Tuna
   B. Butter
   C. Salad oil
   D. Shortening

7. What type of compound is lecithin?
   A. Bile salt
   B. Glycolipid
   C. Lipoprotein
   D. Phospholipid

8. Which of the following is NOT a destination for cholesterol?
   A. Synthesized into bile
   B. Excreted in the feces
   C. Accumulates on walls of veins
   D. Accumulates on walls of arteries

9. In the digestion of fats, emulsifiers function as ________
   A. enzymes.
   B. hormones.
   C. detergents.
   D. chylomicrons.

10. What lipoprotein is responsible for transporting cholesterol back to the liver from the periphery?
    A. Chylomicron
    B. Low-density lipoprotein
    C. High-density lipoprotein
    D. Very-low-density lipoprotein

11. The best energy source for strenuous, aerobic-type activities, such as hiking, is __________
    A. phospholipids.
    B. triglycerides.
    C. carbohydrates.
    D. sterols.

12. Lipoproteins __________
    A. transport lipids through a watery medium, the blood.
    B. are protein-based vehicles for the transport of lipids.
    C. are composed of lipids and proteins.
    D. all of the above.

13. Blood cholesterol is impacted by __________
    A. lifestyle.
    B. nutrition.
    C. heredity.
    D. all of the above.

14. Fat substitutes were developed to __________
    A. help control kcalories in foods.
    B. help people reduce body fat.
    C. make foods taste better.
    D. all of the above.

15. Reducing which of the following foods can reduce triglyceride levels?
    A. Alcohol
    B. Sweets
    C. Complex carbohydrates
    D. A and B

Matching: Select the letter next to the word or phrase that best matches the numbered statements. Answers are used only once.

_____ 16. Common source of *trans*-fatty acids

_____ 17. Good food source of omega-3 fatty acids

_____ 18. A phospholipid

_____ 19. Major dietary precursor for vitamin D synthesis

A. Sardines
B. Chicken
C. Potato chips
D. Cholesterol
E. Lecithin

Essay: Reflect on the item(s) listed below; one or more may be included on the exam.

20. Describe the process of fat hydrogenation, and discuss its advantages and disadvantages.

21. Discuss in detail the digestion, absorption, and transport of dietary lipids, including the sterols.

22. Discuss the composition and function of the major circulating lipoproteins.

ANSWER KEY

The following provides the answers and references for the Practice Test questions.

	Answer	Learning Objective	Reference
1.	C	LO 3	textbook, p. 130
2.	B	LO 4	textbook, p. 131
3.	D	LO 6	textbook, p. 134
4.	A	LO 6	textbook, p. 135
5.	D	LO 5	textbook, p. 132
6.	A	LO 5	textbook, p. 145
7.	D	LO 1	textbook, p. 136
8.	C	LO 7	textbook, p. 137
9.	C	LO 7	textbook, p. 136
10.	C	LO 8	textbook, p. 143
11.	B	LO 2	Video
12.	D	LO 8	Video
13.	D	LO 10	Video
14.	A	LO 11	Video
15.	D	LO 9	Video
16.	C	LO 6	textbook, p. 135
17.	A	LO 5	textbook, p. 132
18.	E	LO 3	textbook, p. 136
19.	D	LO 2	textbook, p. 137
20.		LO 6	textbook, pp. 134–135
21.		LO 7	textbook, pp. 137–144
22.		LO 8	textbook, pp. 141–143

Lesson 6

Fats: Health Effects

LESSON ASSIGNMENTS

Text: Whitney and Rolfes, *Understanding Nutrition*, Chapter 5, "The Lipids: Triglycerides, Phospholipids, and Sterols," pp. 147–164

Video: "Fats: Health Effects" from the series *Nutrition Pathways*

Project: "Diet Analysis Project"

Related Activities:

These activities are not required unless your instructor assigns them. They are offered as suggestions to help you learn more about the material presented in this lesson. Refer to your syllabus to determine which of these activities have been assigned.

Complete the enclosed forms for the following related activities, and return them to your instructor according to the established deadline.

- ❑ "Determining Percentage of Calories and Fat Grams from Fat Intakes"
- ❑ "Calculating a Personal Daily Value for Fat"

Optional Web Activities:

Consult your instructor and/or syllabus for any assigned activities.

OVERVIEW

The role of fat in the diet is important for the normal functioning of the body. What are the consequences of too much fat in the diet? High fat intakes have been linked to cardiovascular disease, diabetes, obesity, and many cancers. This lesson examines the health effects of fatty acid saturation and high dietary cholesterol intake. Furthermore, it examines how blood lipid profiles (which reveal low-density lipoprotein [LDL], triglycerides, total serum cholesterol, and high-density lipoprotein [HDL] levels) are used to detect risks for diseases. Finally, this lesson looks at strategies to improve the health risks associated with high fat intakes.

To complete this lesson, you will answer questions on your personal lipid intake according to directions in the Diet Analysis Project located in the back of this study guide.

LEARNING OBJECTIVES

Upon completing this lesson, you should be able to:

1. Describe the purpose of a blood lipid profile.

2. Describe the health effects of lipids in the body.

3. Cite the recommended intakes of total dietary fat, saturated fat, and dietary cholesterol.

4. Suggest practical ways to reduce total dietary fat, saturated fat, and dietary cholesterol.

5. Suggest ways to balance omega-3 and omega-6 fatty acid intakes.

6. Read food labels to determine fat content of packaged foods, calculate % Daily Value for fat, and state FDA authorized claims regarding fat on labels.

7. Calculate the percentage of kcalories and grams from daily fat intakes.

8. Describe how a traditional meat-based diet of the United States and a traditional olive oil-based diet of the Mediterranean differ and how each can affect health.

9. Provide examples of lower-fat alternatives to high-fat foods when given specific high-fat food choices.

10. Explain the health risks associated with a high-fat diet for someone who has type 2 diabetes.

TEXT FOCUS POINTS

The following focus points are designed to help you get the most from your reading. Review them, then read the assignment. You may want to write notes to reinforce what you have learned.

1. What is the purpose of a blood lipid profile?

2. Name the dietary lipid that raises blood cholesterol even more dramatically than dietary cholesterol does. What type of fat generally raises LDL? What are the risks associated with *trans*-fatty acid intakes? What strategy may be more effective in preventing heart disease even more than reducing total fat intake?

3. How do intakes of monounsaturated fatty acids impact LDL levels in the blood? What are the benefits associated with omega-3 polyunsaturated fatty acid intakes?

4. Explain the effect of dietary fat on the development of the following: cancers (in general), breast cancer, prostate cancer, and obesity.

5. What are the DRI and Dietary Guidelines recommendations for total fat intake, saturated fat intake, and cholesterol? What is the percent of total fat intake for the average person in the United States? What is the percent of total kcalories that comes from saturated fat for the average person in the United States? What is the average dietary cholesterol intake for the average person in the United States?

6. What steps can a person take to reduce the following: total dietary fat, saturated fat, *trans*-fat, and dietary cholesterol. How can most people obtain the right balance between omega-3 and omega-6 fatty acids?

7. What information regarding fat content is optional on food labels? Explain the difference between "% Daily Value" found on food labels and "% kcalories from fat." What are the two health claims approved by the FDA that can appear on food labels with regard to fat?

8. What is the one best change that could be made in the diets of most people? What calculations are necessary to determine the percentage of fat in foods? What calculations are necessary to determine grams of fat in the diet?

VIDEO FOCUS POINTS

The following focus points are designed to help you get the most from the video segment of this lesson. Review them before watching the video. You may want to write notes to reinforce what you have learned.

1. What is the percentage of kcalories from fat recommended by the American Heart Association? What is considered the best type of fat to select? What is the maximum percent of total fat that should be in the form of saturated fat?

2. What steps can be taken to reduce fat in meat-based diets? What health-related problems can occur in people who eat high meat/fat diets? How can a higher fat diet be justified?

3. What is the basis of a Mediterranean diet? What health-related benefits may occur in people who consume a Mediterranean diet? How can olive oil and fish play a role in a healthy diet?

4. Explain how to reduce fat from the following foods: eight ounces of prime rib, mashed potatoes with milk and butter, fried zucchini, garlic bread with butter, a salad with Italian dressing, and a piece of apple pie. How can reducing fat in the diet affect one's overall health?

5. What steps can be taken by a person with type 2 diabetes to reduce fat intake?

PROJECT

This project is required in order to complete the course successfully.

This portion of the lesson is designed to provide you with specific information about nutrient adequacy and balance, number of kcalories consumed, nutrient density, and variety of the foods you typically eat. After analyzing a three-to-seven-day food intake, you will be able to apply the information you have learned to make appropriate changes to improve eating habits.

Refer to the Diet Analysis Project for complete details and directions. Contact your instructor if you have any questions. The Diet Analysis Project is in the back of this student course guide.

Information provided in Chapter 5 will be useful when answering questions pertaining to your Diet Analysis Project with regard to fat intake.

Determining Percentage of Calories and Fat Grams from Fat Intakes

NAME: ______________________________ ID #: ________________________

The purpose of this exercise is to give you practice calculating grams of fat, fat kcalories, and percentage of total kcalories in specific foods. REMEMBER: There are nine kcalories in one gram of fat.

Refer to Chapter 5 in your textbook to help you complete this activity.

PROBLEM #1: Three ounces of pork chop have a total of 275 kcal with 8 fat grams per ounce.

1. ___________ Multiply fat grams per ounce by number of ounces in pork, then multiply by number of kcal in 1 gram of fat to find how many TOTAL fat kcalories are in the pork chop. *(Show your calculations below.)*

2. ___________ Divide total fat kcal (step #1) by total kcal in the pork chop to find the percentage of total kcal from fat. (*Show your calculations below.*)

—Continued on the back

PROBLEM #2: A man consumes 2,500 total kcalories per day and 110 grams of fat.

3. ___________ Multiply grams of fat by number of kcalories in one gram of fat, then divide by the total calories per day to find the percentage of total kcal from fat the man consumes. *(Show your calculations below.)*

PROBLEM #3: A woman consumes 2,200 total kcalories per day and 30 percent of the kcal comes from fat.

4. ___________ Multiply the total kcalories per day by 0.30, then divide by number of kcalories in 1 gram of fat to find the number of grams of fat the woman consumes. *(Show your calculations below.)*

Calculating a Personal Daily Value for Fat

NAME: ______________________________ ID #: ________________________

There are several methods you can use to determine your personal fat allowance. Two methods are described below. The first method multiplies total kcalories by 30 percent then divides by 9 kcalories per gram; the second method deletes the last digit of total kcalories and divides by 3. Practice each method by using total kcalories from two different days of your Diet Analysis Project. *Answers should be the same for each method.*

Refer to Chapter 5 in your textbook for a sample calculation.

METHOD #1:

1. ___________ Day #1 total kcalories.

2. ___________ Multiply the total kcalories for Day #1 by 30 percent, then divide by 9 kcalories per gram. *(Show your calculations below.)*

3. ___________ Day #2 total kcalories.

4. ___________ Multiply the total kcalories for Day #2 by 30 percent, then divide by 9 kcalories per gram. *(Show your calculations below.)*

—Continued on the back

METHOD #2:

5. ____________ Day #1 total kcalories.

6. ____________ Delete the last digit of the total kcalories and divide by 3. (*Show your calculations below.*)

7. ____________ Day #2 total kcalories.

8. ____________ Delete the last digit of the total kcalories and divide by 3. (*Show your calculations below.*)

PRACTICE TEST

The following items will help you check your understanding of this lesson. Compare your answers to the Answer Key at the end of the lesson. Review the course materials related to any incorrect answer.

Multiple Choice: Select the one choice that best answers the question.

1. A LOW risk of cardiovascular disease correlates with high blood levels of ______
   A. triglycerides.
   B. free fatty acids.
   C. high-density lipoproteins.
   D. very-low-density lipoproteins.

2. Among the following, which would be the LEAST effective method to control blood cholesterol levels?
   A. Control body weight.
   B. Eat more insoluble fiber.
   C. Consume less saturated fat.
   D. Exercise intensely and frequently.

3. The results of blood tests that reveal a person's total cholesterol and triglycerides are called a __________
   A. lipid profile.
   B. circulating fat count.
   C. personal lipids count.
   D. degenerative disease assessment.

4. Studies show that diets high in fish oils lead to __________
   A. lower blood pressure.
   B. higher blood cholesterol.
   C. greater tendency of the blood to clot.
   D. decreased storage of omega-3 fatty acids.

5. Which of the following describes a recognized relationship between dietary fat and cancer?
   A. Dietary fat initiates rather than promotes cancer formation.
   B. High saturated fat intake raises the risk for cancer.
   C. Evidence linking fat intake with cancer is stronger than with heart disease.
   D. High intakes of omega-3 fatty acids promote cancer development in animals.

6. According to the *Dietary Guidelines* for Americans, what is the upper limit of fat that should be consumed by a healthy person?
   A. 10%
   B. 20%
   C. 30%
   D. 35%

7. According to the Daily Value, what should be the maximum daily intake of cholesterol?
   A. 50 milligrams
   B. 150 milligrams
   C. 300 milligrams
   D. 1,000 milligrams

8. Which of the following is a drawback of olestra consumption?
   A. It yields 9 kcalories per gram.
   B. It raises blood glucose levels.
   C. It imparts off-flavors to foods.
   D. It inhibits absorption of vitamin E.

9. The American Heart Association recommends the percentage of total calories that come from fat should not exceed __________
   A. 10 percent.
   B. 20 percent.
   C. 30 percent.
   D. 40 percent.

10. A person can reduce fat in a diet that has meat as its base by __________
    A. selecting leaner cuts of meat.
    B. choosing chicken or fish occasionally.
    C. never again eating meat.
    D. both A and B.

11. Some people who eat high-meat diets may have a greater risk for __________
    A. cardiovascular disease.
    B. colon cancer.
    C. obesity.
    D. all of the above.

12. Higher fat intakes can be justified if a person is __________
    A. doing aerobics for short periods of time.
    B. over the age of 55.
    C. performing manual labor for long periods of time.
    D. under the age of 55.

13. The basis of a Mediterranean diet consists of __________
    A. olive oil as the primary fat.
    B. fruits and vegetables as the basis.
    C. lots of grains and breads.
    D. all of the above.

14. Most people who consume a Mediterranean diet usually __________
    A. suffer from obesity.
    B. do not suffer from cardiovascular disease.
    C. show signs of type 2 diabetes.
    D. do not develop lung cancer.

15. If you wanted to reduce the fat and kcalories in a 10-ounce portion of prime rib, you could __________
    A. eat 5 ounces and take the rest home for the next day.
    B. cut away all visible fat but still eat the 10 ounces.
    C. refuse to eat any of the meat.
    D. either A or B.

16. Reducing fat intake for a person with type 2 diabetes can be accomplished by ____
    A. increasing fruits and vegetables.
    B. increasing meat portions.
    C. decreasing breads and grains.
    D. none of the above.

Essay: Reflect on the item(s) listed below; one or more may be included on the exam.

17. Discuss the role of dietary cholesterol and the endogenous production of cholesterol and heart disease. What is meant by "good" and "bad" cholesterol?

18. List strategies for lowering fat intake with minimal impact on diet palatability.

19. Discuss the benefits and possible hazards of dietary fat alternative substances in the diet.

ANSWER KEY

The following provides the answers and references for the Practice Test questions.

	Answer	Learning Objective	Reference
1.	C	LO 1	textbook, p. 143
2.	B	LO 4	textbook, pp. 147–148
3.	A	LO 1	textbook, p. 147
4.	A	LO 5	textbook, pp. 149–150
5.	B	LO 2	textbook, p. 148
6.	D	LO 7	textbook, p. 148
7.	C	LO 6	textbook, p. 148
8.	D	LO 9	textbook, p. 155
9.	C	LO 3	Video
10.	D	LO 8	Video
11.	D	LO 8	Video
12.	C	LO 2	Video
13.	D	LO 8	Video
14.	B	LO 8	Video
15.	A	LO 4	Video
16.	A	LO 10	Video
17.		LO 2	textbook, p. 152
18.		LO 9	textbook, pp. 152–153
19.		LO 2	textbook, pp. 155–156; video

Lesson 7

Protein: Form and Function

LESSON ASSIGNMENTS

Text: Whitney and Rolfes, *Understanding Nutrition*, Chapter 6, "Protein: Amino Acids," pp. 167–187

Video: "Protein: Form and Function" from the series *Nutrition Pathways*

Getting It Together:

The purpose of the "Getting It Together" section is to provide an outline of *new* or *difficult* material presented in the current lesson. This section provides practice in recalling lesson material and utilizes the textbook interactively. Use this section as a learning aid, to reinforce knowledge, or to organize information introduced in the lesson.

Optional Web Activities:

Consult your instructor and/or syllabus for any assigned activities.

OVERVIEW

If you were to ask people, "What is the function of protein in the body?" most would undoubtedly say that protein is used to build strong muscles. The truth is, however, that building muscle is just one of the many varied jobs that protein does in the body, as you will discover. You will begin by examining protein from a chemical standpoint and will look at the differences in amino acids; you will discover which amino acids are essential and which are nonessential and why; and you will look at the structure or conformation of amino acids and see how sequence and folding of the chain impact the functioning of amino acids. As you follow food protein through the GI tract, you will observe what happens when protein is digested, absorbed, and metabolized. Finally, you will examine the impact of protein-energy malnutrition (PEM) on the health and growth of those who suffer from it.

LEARNING OBJECTIVES

Upon completing this lesson, you should be able to:

1. Recognize the basic generic chemical formula of a protein.
2. Define *essential amino acid* and cite which of the amino acids are essential.
3. Explain amino acid sequence and folding in relation to protein function.
4. Describe basic digestion and absorption of protein.
5. Explain what happens during protein synthesis when sequencing errors occur.
6. Describe the roles of protein in the body.
7. Describe how amino acids are metabolized in the body.
8. Describe the characteristics of protein-energy malnutrition (PEM), including its impact on growth and health.
9. Describe the effects of high intakes of animal protein on health.

TEXT FOCUS POINTS

The following focus points are designed to help you get the most from your reading. Review them, then read the assignment. You may want to write notes to reinforce what you have learned.

1. What atoms are found in all proteins, and which one of these atoms is not found in either carbohydrate or fat? What chemical groups or compounds present in amino acids attach to a central carbon? What distinguishes each amino acid from other amino acids?
2. How many amino acids make up proteins? What makes an amino acid "essential"? List the essential amino acids. Define *conditionally essential amino acid.*
3. What links amino acids together? What is the reaction that creates the bond between amino acids? Define the following: *dipeptide*, *tripeptide*, and *polypeptide*. What is responsible for the twisted, tangled shapes of proteins? What enables proteins to perform different tasks in the body? What is denaturation?

4. Where does digestion of protein begin? What happens to protein when it enters the stomach? What is the function of hydrochloric acid? What is pepsin? What happens to polypeptides when they enter the small intestine? Where does absorption of amino acids take place? What is the function of a peptidase? What happens to orally ingested amino acids (predigested protein)?

5. What informs a cell of the amino sequence needed for a specific protein? What is the function of transfer RNA? What can happen if a protein sequence is altered? What is the disease associated with a sequencing error in hemoglobin?

6. To what does "gene expression" refer? Define the following terms: *collagen*, *enzymes*, and *hormones*. Provide two to four examples of hormones and their functions. How does protein help maintain fluid balance? How do proteins respond when the body becomes too acid? What is acidosis? What is alkalosis? What is the function of transport proteins? What is the transport protein that carries oxygen from the lungs to the cells? What transport proteins carry lipids throughout the body? What is the transport protein for iron? What are antibodies? What is an antigen? Summarize the various roles of proteins in the body.

7. What is protein turnover? To what does "amino acid pool" refer? What is nitrogen balance? What is nitrogen equilibrium? Provide examples of people in positive nitrogen balance. Provide examples of people in negative nitrogen balance. What compounds can be made from the amino acid tyrosine? What compounds can be made from tryptophan? Under what condition will amino acids be used for energy? What is deamination? What is the by-product of deamination? What is the fate of ammonia in the body? What happens if a person eats more protein than the body needs?

8. What is protein-energy malnutrition? How would a high protein diet impact heart health? What amino acid is linked to increased risk for heart disease? What is the correlation between protein and bone loss? Are high protein diets a good plan for weight control? What impact does high protein diets have on the kidneys?

VIDEO FOCUS POINTS

The following focus points are designed to help you get the most from the video segment of this lesson. Review them before watching the video. You may want to write notes to reinforce what you have learned.

1. Cite inaccurate statements regarding the function of protein. What is a sound diet for bodybuilders?

2. How does excessive protein intake impact health?

3. What is a typical diet of children with PEM in third-world countries? How does PEM impact health in third-world countries and in the United States?

4. How can federally funded programs in the United States help prevent PEM in school children?

5. What can people do regarding protein intake if they have high blood cholesterol?

GETTING IT TOGETHER

Protein: Form and Function

Refer to Chapter 6 in your textbook for graphic representations of amino acids and complete the outline below. Use the outline as a study aid and to reinforce material.

THE BASICS:

- Basic chemical composition of amino acids:

- Essential amino acids:

- Conditionally essential amino acids:

- Peptide bonds:

- Denaturation:

- Amino acid sequence:

- Roles of proteins:

DIGESTION OF PROTEINS: (Review Chapter 3—"Digestion, Absorption, and Transport")

Refer to Chapter 6 for a graphic representation of protein digestion, and then fill in the blanks below to follow proteins through the GI tract, through absorption, and through metabolism.

1. Hydrolysis of protein begins in the ______________.
2. The loss of a protein's shape and function is known as ______________.
3. The active form of the enzyme pepsinogen is ______________.
4. Polypeptides in the small intestine are hydrolyzed by ______________.
5. Peptides in the small intestine are hydrolyzed by specific ______________.
6. Free amino acids, dipeptides, and tripeptides are absorbed into the ______________ after digestion.
7. Amino acids that can be produced by the body are referred to as ______________ amino acids.
8. The removal of the nitrogen group from an amino acid is known as ______________.
9. ______________ is the toxic by-product of deamination.
10. Ammonia is converted to ______________ in the liver.
11. Urea is excreted as urine from the ______________.

Answer Key to Fill in the Blanks:

1. Stomach
2. Denaturation
3. Pepsin
4. Proteases
5. Peptidases
6. Small intestine
7. Nonessential
8. Deamination
9. Ammonia
10. Urea
11. Kidneys

PRACTICE TEST

The following items will help you check your understanding of this lesson. Compare your answers to the Answer Key at the end of the lesson. Review the course materials related to any incorrect answer.

Multiple Choice: Select the one choice that best answers the question.

1. What element is found in proteins but NOT in carbohydrates and fats?
   A. Carbon
   B. Oxygen
   C. Calcium
   D. Nitrogen

2. What makes protein more complex and unique than carbohydrates or fats?
   A. Central carbon atom
   B. Amino group
   C. Side group
   D. Acid group

3. Which of the following is NOT an essential amino acid in human nutrition?
   A. Proline
   B. Threonine
   C. Methionine
   D. Tryptophan

4. When two amino acids are chemically joined together, the resulting structure is called a __________
   A. dipeptide.
   B. diglyceride.
   C. polypeptide.
   D. polysaccharide.

5. What is meant by the amino acid sequence of a protein?
   A. Number of side chains in the protein
   B. Folding arrangement of the peptide chain
   C. Order of appearance of amino acids in the peptide chain
   D. Order of appearance of only the essential amino acids in the protein

6. Upon eating a hamburger, in what organ is the hydrolysis of its proteins initiated?
   A. Mouth
   B. Stomach
   C. Small intestine
   D. Large intestine

7. In what organ is pepsin active?
   A. Stomach
   B. Pancreas
   C. Small intestine
   D. Large intestine

8. What protein is intimately involved in the formation of scar tissue in wound healing?
   A. Albumin
   B. Thrombin
   C. Collagen
   D. Hydroxyproline

9. What is the relationship between body proteins and water?
   A. Proteins attract water.
   B. Water attracts proteins.
   C. Water degrades proteins.
   D. Proteins form polymers of water.

10. Which of the following does NOT function as a transport protein?
   A. Collagen
   B. Hemoglobin
   C. Sodium-potassium pump
   D. Lipoproteins

11. What term is described by the following quote: “The evil spirit that infects the first child when the second child is born”?
   A. Marasmus
   B. Kwashiorkor
   C. Psychomalnutrition
   D. Postbirth malnutrition

12. Protein is needed for normal development of cells and muscle mass, but excessive amounts of protein __________
   A. are even better and will bring faster results.
   B. will not increase the building of body tissue.
   C. will not add strength to the body.
   D. B and C.

13. Cardiovascular disease, colon cancer, and breast cancer risks have been shown to increase with diets high in __________
   A. all protein.
   B. plant protein.
   C. animal protein.
   D. carbohydrate.

14. Marasmus can be characterized as a condition in which __________
   A. overeating results in bloating and edema.
   B. the person is starving from total malnutrition.
   C. the immune system is impaired, allowing for microbe attack.
   D. B and C.

15. The primary component(s) of the School Lunch Program that help prevent protein-energy malnutrition in U.S. school children is(are) ________
    A. meat or meat alternates.
    B. milk.
    C. pasta and breads.
    D. A and B.

16. For someone who has high blood cholesterol, protein intakes should ________
    A. be less from animal sources and more from plant sources.
    B. represent the bulk of the diet.
    C. not be a concern if they are eating from the Food Pyramid.
    D. A and C.

Matching: Select the letter next to the word or phrase that best matches the numbered statements. Answers are used only once.

_____17. A conditionally essential amino acid

_____18. A large protein that carries oxygen

_____19. The result of protein exposed to severe heat

_____20. A connective tissue protein

A. Collagen
B. Denatured
C. Hemoglobin
D. Albumin
E. Tyrosine

Essay: Reflect on the item(s) listed below; one or more may be included on the exam.

21. Describe the processes involved in cellular protein synthesis. How would synthesis be affected by intake of an otherwise adequate diet which is very low in glycine or low in tryptophan? How would synthesis be affected by a diet that is low in energy?

22. Describe some of the impacts that a high protein diet can have on health.

ANSWER KEY

The following provides the answers and references for the Practice Test questions.

	Answer	Learning Objective	Reference
1.	D	LO 1	p. 168
2.	C	LO 1	p. 168
3.	A	LO 2	p. 168
4.	A	LO 7	p. 169
5.	C	LO 3	pp. 169–170
6.	B	LO 4	p. 171
7.	A	LO 7	p. 171
8.	C	LO 6	p. 175
9.	A	LO 4	p. 176
10.	A	LO 6	p. 177
11.	B	LO 8	Video
12.	D	LO 7	Video
13.	C	LO 9	Video
14.	D	LO 8	Video
15.	D	LO 8	Video
16.	D	LO 9	Video
17.	E	LO 2	p. 169
18.	C	LO 6	p. 170
19.	B	LO 5	p. 170
20.	A	LO 6	p. 175
21.		LO 5	pp. 173–175
22.		LO 8	pp. 182–184

Lesson 8

The Protein Continuum

LESSON ASSIGNMENTS

Text: Whitney and Rolfes, *Understanding Nutrition*, Chapter 6, "Protein: Amino Acids," pp. 181–188 and Appendix D

Video: "The Protein Continuum" from the series *Nutrition Pathways*

Project: Diet Analysis Project

Related Activities:

These activities are not required unless your instructor assigns them. They are offered as suggestions to help you learn more about the material presented in this lesson. Refer to your syllabus to determine which of these activities have been assigned.

Complete the enclosed forms for the following related activities, and return them to your instructor according to the established deadline.

- ❑ "Calculating Personal RDA for Protein"
- ❑ "Plant Proteins for a Meatless Meal"

Optional Web Activities:

Consult your instructor and/or syllabus for any assigned activities.

OVERVIEW

Because foods provide amino acids in differing amounts, how can you tell if the protein you eat has sufficient essential amino acids to support protein synthesis? How can you tell if the protein you eat is high quality, which foods provide the best quality protein, and how much is enough for optimum nutrition? In this lesson, you will discover how protein quality is determined and evaluated by researchers. In conclusion, you will examine the differences between food plans that primarily contain meat protein and food plans that primarily contain plant protein to learn how people can obtain enough high-quality protein from the diet regardless of their protein sources.

LEARNING OBJECTIVES

Upon completing this lesson, you should be able to:

1. Identify the following: limiting amino acids, high quality proteins, complementary proteins, protein digestibility, and reference proteins.

2. Describe how protein quality is measured.

3. Explain the regulations for protein on food labels.

4. Explain the relationship between protein intake and heart disease, cancer, and osteoporosis.

5. Calculate personal RDA for protein, and cite recommended intakes.

6. Describe the effects of protein and amino acid supplements.

7. Differentiate the health effects of vegetarian versus nonvegetarian eating plans.

TEXT FOCUS POINTS

The following focus points are designed to help you get the most from your reading. Review them, then read the assignment. You may want to write notes to reinforce what you have learned.

1. What is a limiting amino acid? What is a high quality dietary protein? What food sources generally provide high quality proteins? To what does "complementary protein" refer? Define protein *digestibility*. What is a reference protein?

2. Refer to Appendix D. Briefly describe the following measures of protein quality: amino acid scoring, biological value, net protein utilization, protein efficiency ratio, and protein digestibility-corrected amino acid score (PDCAAS).

3. Is the % Daily Value for protein required on food labels? What does "% Daily Value" for protein reflect?

4. What is the effect of substituting soy protein for animal protein on high cholesterol? What amino acid does recent research suggest may be a risk factor for heart disease? What does research suggest might be influential factors in raising homocysteine?

What amino acid might prove to be protective against heart disease? What types of cancers may be associated with high-meat diets? How does a high protein intake relate to the development of osteoporosis, obesity, and kidney disease?

5. What are the two reasons the body needs dietary protein? What is the recommended intake of protein as a percent of total kcalories? What is the advice from health experts regarding protein intakes? Why is the daily general RDA for protein intake based on grams per kilogram of body weight? What are the steps necessary to determine individual protein needs (Recommended Daily Allowance [RDA])?

6. How digestible are protein supplements compared to protein foods? What can happen when an excess of single amino acids is eaten? What has research indicated for the amino acids lysine and trytophan with regard to their use and effectiveness for treating specific conditions?

VIDEO FOCUS POINTS

The following focus points are designed to help you get the most from the video segment of this lesson. Review them before watching the video. You may want to write notes to reinforce what you have learned.

1. At what times during life do protein requirements increase? What is the typical protein intake for adults in the United States?

2. Define *meat eaters*. What effects do high-meat diets have on health? What nutrients might be missing from vegetarian diets? What are reasons people choose to become vegetarians? What is a semivegetarian?

3. What are the ways people can reduce daily protein intake? How can reducing protein intake affect weight loss?

PROJECT

This project is required in order to complete the course successfully.

This portion of the lesson is designed to provide you with specific information about nutrient adequacy and balance, number of kcalories consumed, nutrient density, and variety of the foods you typically eat. After analyzing a three-to-seven-day food intake, you will be able to apply the information you have learned to make appropriate changes to improve eating habits.

Refer to the Diet Analysis Project for complete details and directions. Contact your instructor if you have any questions. The Diet Analysis Project is in the back of this student course guide.

Information provided in Chapter 6 will be useful when answering questions pertaining to your Diet Analysis Project with regard to protein intake.

Calculating Personal RDA for Protein

NAME: ______________________________ ID #: ______________________

The following exercise will help you to estimate your personal RDA for protein.

Refer to Chapter 6 in your textbook for a sample calculation.

1. __________ What is your weight in pounds?

2. __________ Convert your weight to kilograms by dividing weight by 2.2. (*Show your calculations below.*)

3. __________ Multiply kilograms by 0.8 g to get your RDA in grams. Males less than eighteen years old, multiply by 0.9 g. (*Show your calculations below.*)

4. __________ What is the RDA for protein for you that is based on age and sex? (Refer to the RDA table in your text.)

—*Continued on the back*

5. Answer the following questions pertaining to your personal protein RDA.

   Cite page numbers in your text to support your answers.

   A. Assume your calculated RDA (in Step 3 above) is different from the RDA stated in the text (in Step 4 above). What would account for that difference?

   B. Explain whether you think the calculated RDA (in Step 3) is adequate for you.

   C. List two examples of protein foods that you could include in your daily eating plan that would offer a high-quality, complete protein WITHOUT cholesterol.
      1)

      2)

Plant Proteins for a Meatless Meal

NAME: ______________________________ ID #: ______________________

This exercise will teach you how to replace animal sources of protein with plant sources of protein to obtain high-quality protein and improve overall nutrient density.

DIRECTIONS: Select one day from your Diet Analysis Project in which you consumed the greatest amount of meat, and then follow the steps below:

1. Substitute ALL meat sources of protein with plant sources of protein. Refer to the chart below for plant combinations for meatless meals.
2. Analyze the plant substitution day using a computer analysis program to determine the following: protein content, kcalories, total fat content, saturated fat content, cholesterol content, and fiber content.
3. Explain the differences you found between the original day and the plant-substituted day.
4. Submit both the original foodlist (with printouts) and the substitute foodlist (with printouts).

Column A plus Column B = A Meatless Protein Meal

(While it is not necessary that you eat the combined foods at the same meal, it is more convenient to do so.)

A	B	Examples of Meatless Meals
Vegetables (1–1½ cups)	Grain/Rice* (2 oz)	1. Cooked vegetables mixed with rice, served with 100 percent whole-wheat bread or crackers
Rice (1–1½ cups)	Beans/ Legumes** (1–1½ cups)	2. Mexican rice and cooked pinto beans, and served with corn tortillas or cornbread 3. Kidney beans and rice added to vegetable soup
Beans/Legumes (1–1½ cups)	Corn (1–1½ cups)	4. Pea soup and cornbread 5. Corn tortillas and steamed pinto beans 6. Kidney beans on tortilla chips broiled with salsa on top 7. Succotash: a mixture of lima beans, corn, and other vegetables
Beans/Legumes (1–1½ cups)	Nuts/Seeds*** (1 tbs)	8. Cooked vegetables with kidney beans, topped with sesame seeds or nuts
Beans/Legumes (1–1½ cups)	Wheat (2 oz)	9. Three-bean salad served with 100 percent whole-wheat bread or crackers 10. Hot vegetables with kidney beans mixed with whole-wheat pasta 11. 100 percent whole-wheat bread with 1 tbs. peanut butter (plus jelly and/or sliced banana for moisture)
Beans/Legumes (1–1½ cups)	Grain/Rice (1–1½ cups)	12. Vegetable soup with pinto or kidney beans (add barley for variety)

*** Grains:** rice, barley, oats, wheat, and corn

**** Legumes:** black, red, or white beans, black-eyed peas, green or yellow peas, garbanzo beans (chick peas), great northern beans, kidney beans, lentils, navy beans, pinto beans, soybeans, and peanuts

***** Nuts/Seeds:** high in protein and high in fat; use sparingly if on a low-fat diet

This table was modified from the original. The source of the table is unknown.

PRACTICE TEST

The following items will help you check your understanding of this lesson. Compare your answers to the Answer Key at the end of the lesson. Review the course materials related to any incorrect answer.

Multiple Choice: Select the one choice that best answers the question.

1. Which of the following is NOT considered to be a source of high quality protein in human nutrition?
   A. Soy
   B. Egg
   C. Corn
   D. Fish

2. What is a "limiting" amino acid in a protein?
   A. A nonessential amino acid present in high amounts which inhibits protein synthesis
   B. An amino acid of the wrong structure to be utilized for protein synthesis efficiently
   C. An essential amino acid present in insufficient quantity for body protein synthesis to take place
   D. An amino acid that limits the absorption of other essential amino acids by competing with them for transport sites within the GI tract

3. Which of the following foods has the best assortment of essential amino acids for the human body?
   A. Eggs
   B. Fish
   C. Corn
   D. Rice

4. Which of the following is a feature of homocysteine?
   A. It is found only in animal foods.
   B. It is a risk factor for osteoporosis.
   C. It is increased in the blood of coffee drinkers.
   D. It is increased in the blood of vitamin C-deficient people.

5. Which of the following is a feature of the protein RDA?
   A. The recommendations are generous.
   B. It is highest proportionately for adult males.
   C. It is established at 8 grams per kilogram of ideal body weight.
   D. An assumption is made that dietary protein is from animal sources only.

6. Which of the following is a known consequence of excess protein intake in animals and human beings?
   A. Increased secretion of water
   B. Decreased secretion of water
   C. Decreased size of the liver and kidneys
   D. Increased protein storage by the liver and kidneys

7. All of the following are advantages of vegetarian diets EXCEPT __________
   A. fat content is lower.
   B. fiber content is higher.
   C. vitamin B_{12} intake is higher.
   D. vitamins A and C are found in liberal quantities.

8. All of the following are features of the Mediterranean diet plan EXCEPT _______
   A. it emphasizes liberal intakes of calcium and iron.
   B. it allows fat intake to rise to 40 percent of total energy.
   C. it includes wine as a means of reducing cardiovascular disease risk.
   D. it places more emphasis on consumption of sweets than red meat.

9. Protein intake needs may increase as a result of illness as well as
   A. when women are lactating.
   B. during moderate physical activity.
   C. from age two to five years.
   D. any of the above.

10. People who consume a diet in which meat is the primary source of protein are
    A. at less risk for cancer than vegetarians.
    B. thought to be at greater risk for cancer than vegetarians.
    C. more likely to develop heart disease than vegetarians.
    D. B and C.

11. Regarding a vegetarian diet, recommendations and observations from experts are that __________
   A. vegan pregnant/lactating women should take a B_{12} supplement.
   B. Lacto-ovo vegetarians might need an iron supplement.
   C. vegetarians eat fewer kcalories and tend to live longer.
   D. all of the above.

12. For people who wish to lose weight, experts might advise them to __________
   A. consider a more vegetarian eating plan for protein choices.
   B. stay at the top of the Food Guide Pyramid for food choices.
   C. focus on the middle of the Food Guide Pyramid for choices.
   D. eat only plant-based proteins, completely eliminating meat.

Matching: Select the letter next to the word or phrase that best matches the numbered statements. Answers are used only once.

______13. A good quality protein source

______14. Combining amino-acid patterns to improve protein quality

______15. An amino acid associated with heart disease

A. Complementary
B. Gelatin
C. Homocysteine
D. Soy

Essay: Reflect on the item(s) listed below; one or more may be included on the exam.

16. Explain the proposed relationship between:
   A. Body homocysteine levels and heart disease
   B. Protein intake and calcium metabolism

17. List the reasons why people take protein supplements. What are the risks involved by taking protein supplements? What population should NOT take supplements?

ANSWER KEY

The following provides the answers and references for the Practice Test questions.

Answer		Learning Objective	Reference
1.	C	LO 1	p. 181
2.	C	LO 1	p. 181
3.	A	LO 2	p. 181
4.	C	LO 4	pp. 182–183
5.	A	LO 3	p. 184
6.	A	LO 6	pp. 183–184
7.	C	LO 7	pp. 182–183
8.	A	LO 7	Video
9.	A	LO 5	Video
10.	D	LO 4	Video
11.	D	LO 7	Video
12.	A	LO 7	Video
13.	D	LO 2	p. 181
14.	A	LO 4	p. 182
15.	C	LO 4	pp. 182–183
16.		LO 4	pp. 182–183
17.		LO 7	pp. 186–187

Lesson 9

Metabolism

LESSON ASSIGNMENTS

Text: Whitney and Rolfes, *Understanding Nutrition*, Chapter 7, "Energy Metabolism:," pp. 197–229. Review metabolism for each of the energy-yielding nutrients found in Chapter 4, Chapter 5, and Chapter 6

Video: "Metabolism" from the series *Nutrition Pathways*

Getting It Together:

The purpose of the "Getting It Together" section is to provide an outline of *new* or *difficult* material presented in the current lesson. This section provides practice in recalling lesson material and utilizes the textbook interactively. Use this section as a learning aid, to reinforce knowledge, or to organize information introduced in the lesson.

Optional Web Activities:

Consult your instructor and/or syllabus for any assigned activities.

OVERVIEW

Now that you know how the food you eat is digested, absorbed, and transported, how does the body actually obtain the energy it needs from that digested food? This is where metabolism comes into play. Metabolism is often the topic of conversation in health clubs and gyms, for body builders, athletes, and others who are interested in controlling weight. Yet, to most people not involved in the biological sciences, it is one of the most confusing—and overwhelming—physiological phenomena associated with the human condition.

You will begin this lesson by looking at the differences between anabolic and catabolic reactions and how these chemical reactions release and produce energy that is transferred by an energy carrier molecule known as ATP. You will follow the basic units of digestion through the metabolic pathways to see how those units become fuels for energy for all the body's activities—immediate and future. The final focus will be on imbalances in energy intake. In this segment, you will examine how food consumed in excess is converted to body fat. Along with feasting, the effects of fasting and starvation on the body will also be examined, as well as the effects of alcohol on metabolic functioning.

LEARNING OBJECTIVES

Upon completing this lesson, you should be able to:

1. Define *metabolism*.
2. Explain the difference between anabolic and catabolic reactions.
3. Describe a coupled reaction.
4. Define *ATP* and explain how the body uses ATP.
5. Explain the function of coenzymes in reactions.
6. Describe the process of glycolysis.
7. Explain how the breakdown products of carbohydrate, fat, and protein digestion proceed through the TCA (Krebs) cycle.
8. Briefly explain amino acid deamination, transamination, and urea synthesis and excretion.
9. Explain how hydrogen with electrons from the TCA cycle enters the electron transport chain (ETC) to produce ATP (energy).
10. Explain why fat provides more energy than carbohydrate or protein.
11. Explain what happens to the metabolism during feasting and during the transition to fasting.
12. Describe what happens to the metabolism during fasting/starvation.
13. Explain the impact of exercise on metabolism, weight gain, and health.
14. Describe the impact of alcohol on the metabolism and nutritional status.

TEXT FOCUS POINTS

The following focus points are designed to help you get the most from your reading. Review them, then read the assignment. You may want to write notes to reinforce what you have learned.

1. Define the following terms: *metabolism*, *anabolism*, *catabolism*, and *coupled reaction*.

2. What is a common high-energy molecule that transfers/carries energy to other compounds? What is the composition of ATP? How is energy derived from ATP? What is the site of metabolic reactions? Where do anaerobic and aerobic reactions occur in the cell?

3. What are coenzymes? What vitamins are used as coenzymes in metabolic reactions? Name the basic units derived from the digestion of food. All compounds that can be converted to pyruvate can be used to make what carbon compound? What compounds cannot be used to make glucose?

4. What is glycolysis? How is oxygen used in the glycolytic pathway? To what compound can a glucose molecule convert? What happens to pyruvate if oxygen is available? What happens to pyruvate when there is not enough oxygen available? When do lactic acid reactions occur? What is the Cori cycle, and where does it take place?

5. What step is irreversible in metabolic pathways? What happens to acetyl CoA if a cell needs energy? What happens to acetyl CoA if a cell does not need energy? What is the TCA cycle? What compound cannot be made from acetyl CoA?

6. What is the fate of acetyl CoA from fatty acid conversion? Briefly describe the process known as "fatty acid oxidation." What compound cannot be formed from fatty acids? How much of a fat molecule can be used to make glucose? What part of a fat molecule can be used to make glucose?

7. What is the first thing to happen to amino acids if they are needed for energy or eaten in excess? What is the fate of amino acids that convert to pyruvate? What happens to amino acids that are converted to acetyl CoA if the body does not need energy?

8. What are the products of amino acid deamination? What is transamination? What can be synthesized from some of the ammonia in a deamination reaction? Where is urea formed? What is the fate of urea? What is the risk associated with a high-protein intake? What nutrient is required in greater amounts if one eats a high-protein diet?

9. What is the first 4-carbon compound to enter the TCA cycle? Why is oxaloacetate critical in the TCA cycle? From what dietary source is oxaloacetate primarily made? As acetyl CoA breaks down to carbon dioxide in the TCA cycle, what atoms are removed from the TCA compounds and enter the electron transport chain (ETC)? How is hydrogen with the electrons from the TCA cycle transferred to the ETC? What happens to electron carriers as they pass electrons down the ETC? What is produced as electrons are passed down the ETC? What does the very last step in the ETC produce?

10. Which of the energy-yielding nutrients provides the most energy (ATP)? Upon what atom is the ultimate number of ATP based? What is the body's preferred energy-storage form?

11. Briefly explain why dietary fat, but not dietary carbohydrate, is converted to adipose tissue more efficiently. What happens to excess dietary protein if it is not needed for building tissues or used for energy? What is the first compound stored from excess carbohydrate intake? How do carbohydrate intakes impact glucose oxidation? What happens when glycogen stores are filled? What happens to excess dietary fat if the body doesn't need more energy?

12. At the beginning of a fast, what fuel(s) is (are) most depleted? What is the preferred fuel for the brain, red blood cells, and the nervous system? What amount of the body's total glucose is used by the brain and nerves? What percent of the body's glucose is provided by protein during the first few days of a fast? What percent is provided by fat?

13. After several days of fasting, what is the brain's alternate fuel source? What is ketosis? What happens to appetite during a fast? What happens to the metabolism during a fast? What happens to lean body losses and fat losses during a fast? List the physical symptoms of starvation.

VIDEO FOCUS POINTS

The following focus points are designed to help you get the most from the video segment of this lesson. Review them before watching the video. You may want to write notes to reinforce what you have learned.

1. How is metabolism like a power plant?
2. What nutrient on a weight-to-weight basis is the most calorie dense?
3. How can physical activity affect calorie expenditure and metabolism? How does extra body fat affect the circulatory system?
4. What is the effect of fasting on the brain and heart?
5. How does alcohol affect metabolism?

GETTING IT TOGETHER

Understanding Energy Metabolism

Because the topic of metabolism can be rather daunting for students, this Getting It Together section is more detailed than previous ones and is designed to help you better understand how the body produces energy in the form of ATP. By learning about energy metabolism, you will also learn how any food you eat *in excess of the body's needs* will be stored as body fat (and can increase blood glucose, as well). The section begins with dietary carbohydrate and follows its breakdown product, glucose, through glycolysis, the TCA cycle, and finally, through the electron transport chain to energy (ATP) production. It also follows the breakdown products of dietary fat (glycerol and fatty acids) and dietary protein (amino acids) through the same pathways.

Review the graphic, "METABOLISM: How Food Provides Energy (ATP)," shown on the next page, and then fill in the blanks that follow. You may also refer to Chapter 7 in your textbook for additional graphics representing metabolism. Use this section as a study aid and to reinforce material.

One-Way Reaction: → **Two-Way Reaction:** ↔

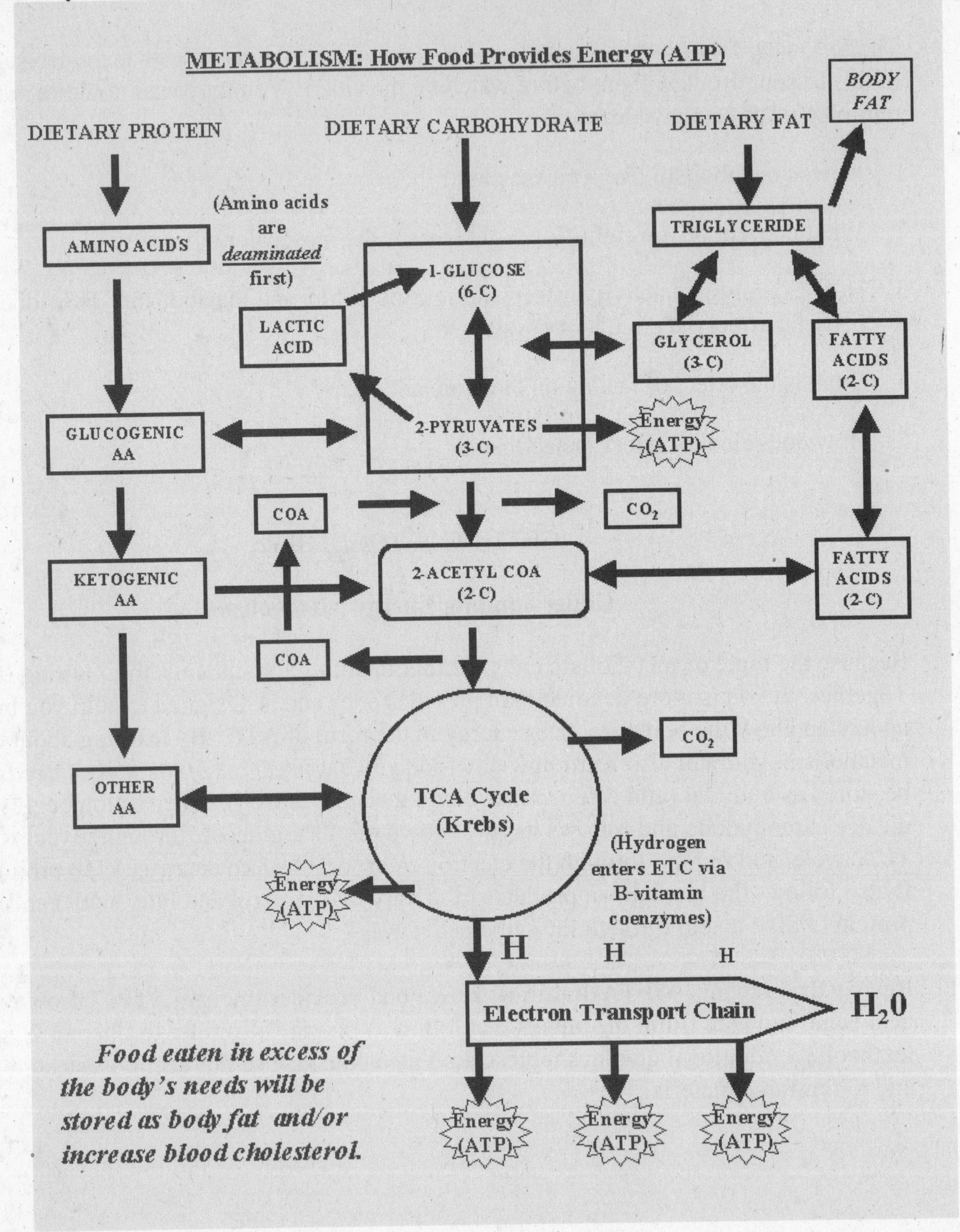
METABOLISM: How Food Provides Energy (ATP)
BODY FAT
DIETARY PROTEIN
DIETARY CARBOHYDRATE
DIETARY FAT
(Amino acids are deaminated first)
AMINO ACIDS
TRIGLYCERIDE
1-GLUCOSE (6-C)
LACTIC ACID
GLYCEROL (3-C)
FATTY ACIDS (2-C)
GLUCOGENIC AA
2-PYRUVATES (3-C)
Energy (ATP)
COA
CO2
KETOGENIC AA
2-ACETYL COA (2-C)
FATTY ACIDS (2-C)
COA
CO2
OTHER AA
TCA Cycle (Krebs)
(Hydrogen enters ETC via B-vitamin coenzymes)
Energy (ATP)
H
H
H
Electron Transport Chain
H20
Food eaten in excess of the body's needs will be stored as body fat and/or increase blood cholesterol.
Energy (ATP)
Energy (ATP)
Energy (ATP)

Glucose in the Energy Pathways

A. GLYCOLYSIS: The breakdown of glucose to pyruvate to energy

1. There are ______________ carbons in a glucose molecule.
2. One glucose molecule will produce ______________ pyruvate molecules consisting of ______________ carbons each.
3. The metabolic pathway that breaks down glucose to pyruvate is known as ______________.
4. Glycolysis occurs in the ______________ of the cell and the process is ______________, which means "without oxygen."
5. Limited energy in the form of ______________ is produced during glycolysis.

B. THE CORI CYCLE: An alternate anaerobic pathway to energy production

1. If enough oxygen is NOT available, especially during high-intensity exercise, pyruvate converts to ______________.
2. Lactic acid accumulates in the ______________ due to limited availability of oxygen and limited clearance of ______________.
3. When muscles relax, circulating blood carries lactic acid to the ______________, which will then convert it to ______________.
4. The conversion of lactic acid to glucose is known as the ______________.

C. THE TRICARBOXYLIC ACID (TCA) CYCLE: The aerobic pathway to energy production

1. If enough oxygen is available to the cell, pyruvate converts to a 2-carbon compound known as ______________.
2. The "CoA" part of acetyl CoA is a coenzyme that comes from the B vitamin, ______________.

3. If enough oxygen is available to the cell and the cell needs more ATP (energy), acetyl CoA enters the ________________.

4. Other names for the TCA cycle include ________________ and ________________.

5. The TCA cycle occurs in the ________________ of the cell and the process is ________________, which means "with oxygen."

6. If pyruvate continues through the energy pathway, two of the three carbons in pyruvate combine with coenzyme A to become ________________ and the third carbon is expelled as ________________.

7. Pyruvate *can* convert to ________________ or ________________, but acetyl CoA *cannot* convert to ________________ or ________________.

8. Acetyl CoA will proceed through the TCA cycle ONLY if the cell needs more ________________.

9. Acetyl CoA that does not enter the TCA cycle will be stored as ________________ and/or increase ________________.

10. When acetyl CoA enters the TCA cycle it converts to several other compounds that produce ________________, ________________, and provides ________________ atoms.

11. The TCA cycle provides ________________ atoms for the electronic transport chain to produce ________________ and ________________.

D. THE ELECTRON TRANSPORT CHAIN (ETC): The pathway to abundant energy production

1. When acetyl CoA converts to other compounds in the TCA cycle, ________________ atoms with their energetic ________________ are removed from those compounds.

2. The ETC occurs in the ________________ of the cell and the process is ________________, which means "with oxygen."

3. The transfer of hydrogen to the ETC is made possible because of ________________ made from B vitamins.

4. As hydrogen with electrons is passed down the ETC, it releases energy which is captured in _______________.

5. At the end of the ETC, very low-energy hydrogen atoms combine with oxygen to form _______________.

6. The number of _______________ atoms that enter the ETC determine the amount of _______________ that will be produced.

Fat in the Energy Pathways

1. When dietary fat is needed for energy, the 3-carbon backbone of the triglyceride molecule, known as _______________, enters the energy pathway between the compounds _______________ and _______________.

2. The _______________ portions of a triglyceride are taken apart, 2 carbons at a time, through a process known as _______________.

3. A fatty acid that contains 20 carbons will yield _______________ acetyl CoA molecules.

4. The 2-carbon fragments of a fatty acid can enter the energy pathway only as _______________ compounds that can enter the TCA cycle ONLY if the cell needs more _______________.

5. Acetyl CoA from triglyceride that is not used for ATP (energy) will be stored as _______________ and/or increase _______________.

6. If acetyl CoA from triglyceride enters the TCA cycle, the process continues through the _______________ to produce abundant ATP (energy) and water.

7. When a triglyceride molecule breaks down into its basic units, _______________ *can* be used to make glucose, but _______________ *cannot* be used to make glucose.

Protein in the Energy Pathways

1. Before protein can enter the energy pathway, enzymes must first ______________ (remove nitrogen groups) the amino acids.

2. The process of deamination produces the compound ______________ that can then be used to make other amino acids or excreted as ______________.

3. An amino acid that can enter the energy pathway between glucose and pyruvate is referred to as a ______________ amino acid.

4. An amino acid that can enter the energy pathway as acetyl CoA is referred to as a ______________ amino acid.

5. Amino acids that convert to pyruvate can "go up" to make ______________ or "go down" to make ______________.

6. Amino acids that convert to acetyl CoA will enter the TCA cycle ONLY if the cell needs more ______________.

7. Amino acids that convert to acetyl CoA cannot be used to make ______________.

8. After acetyl CoA enters the TCA cycle, the process continues on through the ______________ as described previously for dietary carbohydrate and fat.

9. Just like dietary carbohydrate and fat that is used for energy, dietary protein can also provide ______________ atoms that will enter the electron transport chain to produce ______________.

10. If ATP (energy) is NOT needed from amino acids, they will produce ______________ and/or increase ______________.

Summary of Energy Metabolism

1. All dietary carbohydrate, fat, and protein can break down to the common compound _______________ that can enter the TCA cycle.

2. Acetyl CoA enters the TCA cycle only if the cell needs more _______________.

3. Most of the ATP (energy) made is produced in the part of the energy pathway known as the _______________.

4. Abundant ATP (energy) comes from dietary fat because fat provides more _______________.

5. The more hydrogen atoms provided, the more _______________ will be produced.

6. If eaten in excess of the body's needs, all dietary carbohydrate, fat, and protein will produce _______________ and/or increase _______________ levels.

Answer Key to Fill in the Blanks

GLUCOSE IN THE ENERGY PATHWAYS

A. GLYCOLYSIS
1. Six
2. Two; three
3. Glycolysis
4. Cytoplasm; anaerobic
5. ATP (adenosine triphosphate)

B. THE CORI CYCLE
1. Lactic acid
2. Muscles; carbon dioxide
3. Liver; glucose
4. Cori cycle

C. THE TRICARBOXYLIC ACID (TCA) CYCLE

1. Acetyl CoA
2. Pantothenic acid
3. TCA cycle
4. Krebs cycle; Citric acid cycle
5. Mitochondria; aerobic
6. Acetyl CoA; carbon dioxide
7. Glucose; acetyl CoA; pyruvate; glucose
8. ATP (energy)
9. Body fat (adipose tissue); blood cholesterol
10. Carbon dioxide; some ATP; hydrogen
11. Hydrogen; ATP (energy); water

D. THE ELECTRON TRANSPORT CHAIN (ETC)

1. Hydrogen; electrons
2. Mitochondria; aerobic
3. Coenzymes
4. ATP (phosphate bonds)
5. Water
6. Hydrogen; ATP (energy)

FAT IN THE ENERGY PATHWAYS

1. Glycerol; pyruvate; glucose
2. Fatty acid; fatty acid oxidation
3. 10
4. Acetyl CoA; ATP (energy)
5. Body fat (adipose tissue); blood cholesterol
6. Electron transport chain
7. Glycerol; fatty acids

PROTEIN IN THE ENERGY PATHWAYS

1. Deaminate
2. Ammonia; urine
3. Glucogenic
4. Ketogenic
5. Glucose; acetyl CoA
6. ATP (energy)
7. Glucose
8. Electron transport chain
9. Hydrogen; ATP (energy)
10. Body fat; cholesterol

SUMMARY OF ENERGY METABOLISM

1. Acetyl CoA
2. ATP (energy)
3. Electron transport chain
4. Hydrogen
5. ATP (energy)
6. Body fat (adipose tissue); blood cholesterol

PRACTICE TEST

The following items will help you check your understanding of this lesson. Compare your answers to the Answer Key at the end of the lesson. Review the course materials related to any incorrect answer.

Multiple Choice: Select the one choice that best answers the question.

1. Which of the following describes the sum of all chemical reactions that go on in living cells?
   A. Digestion
   B. Metabolism
   C. Absorption
   D. Catabolism

2. What is the major energy carrier molecule in most cells?
   A. ATP
   B. Glucose
   C. Pyruvate
   D. A kcalorie

3. Glycolysis is the conversion of __________
   A. glycogen to fat.
   B. glucose to glycogen.
   C. glucose to pyruvate.
   D. glycogen to protein.

4. Which of the following CANNOT be formed from acetyl CoA molecules?
   A. Glucose
   B. Cholesterol
   C. Stearic acid
   D. Carbon dioxide

5. The Cori cycle involves the conversion of __________
   A. lactic acid and glucose.
   B. glucose and amino acids.
   C. pyruvate and citric acids.
   D. fatty acids and acetyl CoA.

6. In a triglyceride that contains 54 carbon atoms, how many can become part of glucose?
   A. 3
   B. 9
   C. 54
   D. 108

7. When protein consumption is in excess of body needs and energy needs are met, the excess amino acids are metabolized and the energy in the molecules is _______
   A. stored as fat only.
   B. excreted in the feces.
   C. stored as amino acids only.
   D. stored as glycogen and fat.

8. What is the most likely reason for having an abnormally high blood urea level?
   A. Liver dysfunction
   B. Kidney dysfunction
   C. Protein intake of twice the RDA
   D. Protein intake of one-tenth the RDA

9. After the first day or so of fasting, which of the following is most depleted in the body?
   A. Glycogen
   B. Fatty acids
   C. Amino acids
   D. Triglycerides

10. How soon would death occur from starvation if the body was unable to shift to a state of ketosis?
    A. Within three weeks
    B. Less than two weeks
    C. Between five and six weeks
    D. Between two and three months

11. Which of the following dietary nutrients would most *rapidly* reverse a state of ketosis in a starving person?
    A. Fat
    B. Protein
    C. Amino acids
    D. Carbohydrate

12. A power plant and the body's metabolism are similar in that both __________
    A. generate energy.
    B. store unused energy.
    C. release energy when needed.
    D. all of the above.

13. Every time you put on an extra pound of body fat, you can __________
    A. increase blood pressure.
    B. decrease blood pressure.
    C. become more susceptible to some cancers.
    D. A and C.

Fill in the Blank: Insert the correct word or words in the blank for each item.

14. The nutrient that is the most calorie dense on a weight-to-weight basis is _______________.

15. Exercise has been shown to _________________ metabolism.

16. Alcohol intake impacts the body's use of B vitamins by _________________ ______________________.

Matching: Select the letter next to the word or phrase that best matches the numbered statements. Answers are used only once.

_____17. Example of a catabolic reaction

_____18. An irreversible reaction

_____19. A product of deamination

_____20. A small nonprotein organic substance that promotes optimal activity of an enzyme

A. Coenzyme
B. Synthesis of pyruvate from glycogen
C. Synthesis of acetyl CoA from glucose
D. Ammonia
E. Urea

Essay: Reflect on the item(s) listed below; one or more may be included on the exam.

21. What are the major differences between aerobic and anaerobic metabolism? Give an example of an aerobic reaction and an anaerobic reaction.

22. Discuss ways in which the body's metabolism adapts to conditions of fasting/starvation. How do these adaptations affect the rate of weight loss when dieting?

23. What is ketosis and how can it be identified? What conditions typically induce a state of ketosis? What are the adverse effects of this abnormality?

ANSWER KEY

The following provides the answers and references for the Practice Test questions.

	Answer	Learning Objective	Reference
1.	B	LO 1	p. 198
2.	A	LO 3, 4	p. 199
3.	C	LO 6	p. 203
4.	A	LO 7	p. 202
5.	A	LO 2	p. 206
6.	A	LO 9	p. 207
7.	D	LO 11	pp. 207–208
8.	B	LO 11	pp. 207–208
9.	A	LO 12	p. 217
10.	A	LO 12	p. 217
11.	D	LO 12	p. 218
12.	D	LO 1	Video
13.	D	LO 11	Video
14.	fat (fatty acids)	LO 10	Video
15.	increase	LO 13	Video
16.	preventing absorption (depletes vitamins)	LO 14	Video
17.	B	LO 2	p. 199
18.	C	LO 7	p. 206
19.	D	LO 8	p. 207
20.	A	LO 5	p. 200
21.		LO 2	pp. 205–206
22.		LO 11	pp. 215–218
23.		LO 10	pp. 215–218

Lesson 10

Weight Control: Energy Regulation

LESSON ASSIGNMENTS

Text: Whitney and Rolfes, *Understanding Nutrition*, Chapter 8, "Energy Balance and Body Composition," pp. 231–258

Video: "Weight Control: Energy Regulation," from the series *Nutrition Pathways*

Related Activities:

These activities are not required unless your instructor assigns them. They are offered as suggestions to help you learn more about the material presented in this lesson. Refer to your syllabus to determine which of these activities have been assigned.

Complete the enclosed forms for the following related activities, and return them to your instructor according to the established deadline.

❒ "Estimating Personal Energy Expenditure"
❒ "Calculating Personal Body Mass Index"
❒ "Assessing Body Weight and Fat Distribution to Determine Risk for Disease"

Optional Web Activities:

Consult your instructor and/or syllabus for any assigned activities.

OVERVIEW

As you learned in the metabolism lesson, the food you eat is not only converted to energy for the present but also converted to energy for future needs—mainly in the form of body fat which has a virtually unlimited capacity for storage. It seems amazing, considering the daily intake of food and the daily expenditure of energy from that food, that most of us are able to maintain a fairly stable body weight from year to year.

This lesson delves into the complexities of weight control by discussing energy balance, body weight, body composition, and the ways health is affected by over- and underweight. You will begin by looking at how the body expends the energy it obtains from food through basal metabolism, physical activity, thermic effect, and adaptive thermogenesis. You will estimate your personal energy requirements and begin to understand how your own metabolism affects weight gain, loss, or maintenance. The importance of healthy body weight versus the distorted view of the "ideal" body presented by the media is discussed in this lesson as well. Finally, you will learn the best ways to determine healthy body composition and the factors for good health that must be taken into account beyond a bathroom scale.

LEARNING OBJECTIVES

Upon completing this lesson, you should be able to:

1. Describe energy balance.
2. Describe how energy is measured in food.
3. Identify the factors that affect food intake.
4. Define *thermogenesis*.
5. Describe the components of energy expenditure and factors that affect them.
6. Calculate estimated energy requirement.
7. Define "healthy" and "desirable" body weight.
8. Describe the methods of determining body weight.
9. Describe the methods of determining body composition.
10. State the health risks associated with being under- and overweight.
11. Describe the influence of the media on body image.
12. Explain the differences between anorexia nervosa and bulimia nervosa.

TEXT FOCUS POINTS

The following focus points are designed to help you get the most from your reading. Review them, then read the assignment. You may want to write notes to reinforce what you have learned.

1. How many kcal are in one pound of body fat? How many kcal per day must a person expend (or take in) to lose (or gain) one pound of weight in a month?

2. What information does a bomb calorimeter provide? What do *direct* and *indirect* calorimetry measure?

3. What is the difference between hunger and appetite? What is the difference between satiation and satiety? Cite examples of eating habits that override the signals of hunger and satiety. Of the energy-yielding nutrients, which one is the most satiating? What organ is the control center for eating behavior? What is neuropeptide Y?

4. What is thermogenesis? What are the three main categories of thermogenesis? Name a fourth category of thermogenesis that is sometimes involved in energy expenditure.

5. How much energy spent in a day comes from basic metabolic functions? What factors raise basic metabolic rate (BMR)? What factors lower BMR? What three factors affect energy needs during physical activity? How does body weight affect energy expenditure? How does the activity's duration, frequency, and intensity affect energy expenditure? Define the *thermic effect of food*. How much energy per day comes from the thermic effect of food? Define *adaptive thermogenesis*.

6. What components of energy expenditure are considered when calculating estimated energy requirement? To estimate energy spent on BMR, what factors are used? What steps are taken to estimate energy output for an individual?

7. What three criteria are used to define "healthy" body weight?

8. Define *body mass index* (BMI). What BMI is associated with overweight and obesity? What BMI is associated with the lowest risk to health for younger and older adults? What two pieces of information does BMI fail to measure? Based on BMI, how is ideal body weight determined? What is a reasonable target for most overweight individuals?

9. What is the percent body fat range for a normal-weight man and for a normal-weight woman? What is one important criterion for determining how much a person should weigh or how much body fat a person needs? At what percent body fat do health problems generally develop in men and women whose ages are below and above forty years? How does fat distribution relate to increased risk for disease? What is "central" obesity? What is meant by "apples" and "pears" relative to fat distribution? What is the most valuable and practical indicator of fat distribution and abdominal fat? How many inches around the waist indicates higher risk for obesity-related problems?

10. Briefly describe three methods used to assess body composition.

11. What health risks are associated with being underweight? How many people die annually from obesity-related diseases? Explain the relationship between overweight and the following: cardiovascular disease, type 2 diabetes, inflammation and the metabolic syndrome, and cancer.

12. Explain the three components of the female athlete triad. Describe the differences between anorexia and bulimia with regard to the following: cause, diagnosis, symptoms, and treatment. Differentiate between bulimia and binge-eating disorder. What are the criteria for diagnosis of binge-eating disorder?

VIDEO FOCUS POINTS

The following focus points are designed to help you get the most from the video segment of this lesson. Review them before watching the video. You may want to write notes to reinforce what you have learned.

1. Identify factors that affect food intake other than appetite and hunger. What should internal cues regarding food intake tell people?

2. Who is the primary target of media with regard to ideal body image and size? How has the media contributed to the dieting craze in the United States?

3. Define "desirable" body weight. What is good advice regarding a person's body size?

4. What factor has greatest influence on maintaining weight loss?

5. How does a lower metabolic rate impact ability to lose weight successfully?

Estimating Personal Energy Expenditure

NAME: ______________________________ ID #: ________________________

This exercise will help you to estimate the number of kcalories per day you need to maintain your weight.

See "How to Estimate Energy Requirements" in Chapter 8 of your textbook for an example.

1. ___________ Your current (or preferred) weight in pounds.

2. ___________ Convert pounds to kilograms by dividing your body weight by 2.2. (*Show your calculations below.*)

3. ___________ Convert height in inches to meters by dividing inches by 39.37. *(Show your calculations below.)*

TOTAL DAILY ENERGY NEEDS (represents kcalories needed to maintain the weight indicated in Step 1)

4. ___________ Refer to page 240 to estimate total kcalories per day. (*Fill in the formula below.*)

   Men: 662 – 9.53 x age + PA x ([15.91 x wt] + [539.6 x ht])

   Women: 354 – 6.91 x age + PA x ([9.36 x wt] + [726 x ht])

5. Do you feel the above estimation accurately reflects the number of kcalories per day you need to maintain the weight indicated in Step 1? Why or why not? Cite page numbers in text to support your answer.

Calculating Personal Body Mass Index

NAME: ______________________________ ID #: ______________________

This activity will help you to estimate your Body Mass Index (BMI).

Refer to Chapter 8 of your textbook to complete this activity.

1. __________ Your height in inches.

2. __________ Your height in meters. Divide your height in inches by 39.37 to obtain your height in meters. (*Show your calculations below.*)

3. __________ Your weight in pounds.

4. __________ Your weight in kilograms. Divide your weight in pounds by 2.2 to obtain your weight in kilograms. (*Show your calculations below.*)

5. __________ Your BMI. Divide your weight in kilograms by the square of your height in meters: BMI = kg ÷ $(m)^2$. (*Show your calculations below.*)

6. What are the advantages and disadvantages associated with BMI measurements? Cite page numbers in the text to support your answer.

   Advantages:

   Disadvantages:

Assessing Body Weight and Fat Distribution to Determine Risk for Disease

NAME: ______________________________ ID #: ______________________

The function of this activity is to help you determine whether your current weight and body composition puts you at risk for CVD, stroke, hypertension, cancer, and/or diabetes.

Refer to Chapter 8 for sample calculations.

DIRECTIONS: Complete the information in the box below, then respond to the questions that follow.

Age:	________	Sex (M or F):	________
Height (inches):	________	Height (meters) (meters = inches ÷ 39.37):	________
Current weight (pounds):	________	Weight (kilograms) (kilograms = lbs ÷ 2.2):	________
Waist (inches at navel):	________	Hips (inches at widest part):	________

1. ____________ Your Body Mass Index: BMI = kilograms ÷ $(\text{meters})^2$. (Show calculations below.)

2. According to your BMI, you are (check one):

_____	less than 20	=	underweight
_____	20–25	=	normal (very low risk of disease)
_____	25–30	=	overweight (low risk of disease)
_____	greater than 30	=	obese (moderate to very high risk)

—Continued on the back

3. ____________ Your waist-to-hip ratio (WHR ratio = waist in inches ÷ hips in inches. Show calculations below.) (See Appendix E-14)

4. According to your WHR (Step 6), you are (check one):

Females:

___ less than 0.8 = low risk of disease

___ greater than 0.8 = high risk of disease

Males:

___ less than 0.95 = low risk of disease

___ greater than 0.95 = high risk of disease

5. Given your *genetic background* and the calculations in Steps1–4 above, explain whether you are at risk for developing heart disease, cancer, or diabetes. (Cite page numbers in the text to support your answers.)

6. Consider the knowledge you have gained from this activity and from the lessons on weight control. Now, disregard the image of the "ideal" body as portrayed by the media. Do you believe you should lose weight, and if so, how much weight should you lose?

PRACTICE TEST

The following items will help you check your understanding of this lesson. Compare your answers to the Answer Key at the end of the lesson. Review the course materials related to any incorrect answer.

Multiple Choice: Select the one choice that best answers the question.

1. What would be the approximate weight gain of a person who consumes an excess of 500 kcalories daily for one month?
   A. Half of a pound
   B. Two pounds
   C. Three pounds
   D. Four pounds

2. What instrument is used to measure the energy content of foods?
   A. Energy chamber
   B. Exothermic meter
   C. Bomb calorimeter
   D. Combustion chamber

3. External cues that may cause an obese person to respond helplessly to food typically include all of the following EXCEPT ________
   A. TV commercials.
   B. outdoor exercises.
   C. availability of food.
   D. "time of day" patterns.

4. Which of the following is a feature of the basal metabolic rate (BMR)?
   A. Fever decreases the BMR.
   B. Fasting increases the BMR.
   C. Pregnancy increases the BMR.
   D. Females have a higher BMR than males on a body weight basis.

5. If a dancer and a typist are the same height and have the exact same body build, the dancer will be heavier because she has __________
   A. more body fat.
   B. stronger bones.
   C. stronger muscles.
   D. more muscle mass.

6. Which of the following describes a feature of body weight standards?
   A. Current weight-for-height tables specify recommendations by age and sex.
   B. Weight-for-height tables for adults have been the same for at least fifteen years.
   C. The weight-for-height tables are used less frequently than body mass index.
   D. Weight-for-height tables and body mass index have been blended to yield a more precise measure known as the Minimal Health Risk classification.

7. Which of the following is a characteristic associated with using weight measures to assess risk of disease?
   A. They are expensive to perform.
   B. They are complicated to perform.
   C. They are subject to inaccuracies.
   D. They fail to quantitate total body fat.

8. Which of the following defines central obesity?
   A. Accumulation of fat during the mid-years of life
   B. Storage of excess fat around the central part of the body
   C. Overfatness due to a large number of interacting behavioral problems
   D. Overfatness due to reliance on high-fat foods as a central part of the diet

9. What is the major factor that determines metabolic rate?
   A. Age
   B. Gender
   C. Amount of fat tissue
   D. Amount of lean body tissue

10. Factors that affect food intake include not only hunger and appetite but also __________
    A. genetics.
    B. the drive for energy (kcalories).
    C. age.
    D. all of the above.

11. What population has been the primary target of body image idealization especially in the American culture?
    A. Males
    B. Adolescents
    C. Females
    D. Children

12. How has the media contributed to the dieting phenomenon in the United States?
    A. By idealizing the average housewife
    B. By idealizing the average athlete
    C. By idealizing female models
    D. By idealizing male models

13. Desirable weight might be defined as that weight which __________
    A. is not associated with hypertension, diabetes, or heart disease.
    B. is associated with the lowest mortality (death) rate.
    C. can be maintained for more than five years.
    D. A and B.

14. In studies performed on women who were able to maintain their lost weight, findings showed that __________
    A. they attended regular support groups.
    B. they kept regular track of food intake.
    C. 90 percent of them exercised regularly.
    D. all of the above.

15. If people are trying to lose weight and are 20 percent below the predicted equation for size, sex, and age for metabolic rate, they will __________
    A. have a harder time burning kcalories and not lose weight.
    B. be tired most of the time and gain weight.
    C. expect to lose weight quickly on 1,500 kcalories per day.
    D. A and B.

Matching: Select the letter next to the word or phrase that best matches the numbered statements. Answers are used only once.

_____ 16. Approximate number of kcalories in two pounds of body fat

_____ 17. A psychological desire to eat

_____ 18. Eating in response to the time of day

_____ 19. Term that describes the energy needed to process food

_____ 20. Synonymous with the thermic effect of food

A. Appetite
B. External cue
C. 7,000
D. Thermic effect
E. Specific dynamic effect
F. 8,000

Essay: Reflect on the item(s) listed below; one or more may be included on the exam.

21. List the major components that contribute to the body's daily expenditure of energy. Compare the relative contributions of a sedentary person with a marathon runner of the same body weight.

22. Explain the adverse effects of excess body fat deposited around the abdominal region.

ANSWER KEY

The following provides the answers and references for the Practice Test questions.

Answer	Learning Objective	Reference
1. D	LO 1	p. 232
2. C	LO 2	p. 232
3. B	LO 3	pp. 233–234
4. C	LO 5	p. 237
5. D	LO 9	p. 243
6. C	LO 8	pp. 242–243
7. D	LO 8	pp. 242–243
8. B	LO 9	p. 244
9. D	LO 5	pp. 236–237
10. D	LO 3	Video
11. C	LO 12	Video
12. C	LO 12	Video
13. D	LO 7	Video
14. C	LO 5	Video
15. A	LO 10	Video
16. C	LO 1	p. 232
17. A	LO 3	p. 233
18. B	LO 5	pp. 233–234
19. D	LO 4	p. 238
20. E	LO 4	p. 238
21.	LO 5	pp. 236–238
22.	LO 7	pp. 244–249

Lesson 11

Weight Control: Health Effects

LESSON ASSIGNMENTS

Text: Whitney and Rolfes, *Understanding Nutrition*, Chapter 9, "Weight Management: Overweight, Obesity, and Underweight," pp. 261–295

Video: "Weight Control: Health Effects" from the series *Nutrition Pathways*

Project: "Diet Analysis Project"

Related Activity:

This activity is not required unless your instructor assigns it. It is offered as a suggestion to help you learn more about the material presented in this lesson. Refer to your syllabus to determine whether this activity has been assigned.

Complete the attached form for the following related activity, and return it to your instructor according to the established deadline.

- ❐ "Designing a Personal Weight Control Plan"
- ❐ "Evaluating a Weight-Loss Program or Product"

Optional Web Activities:

Consult your instructor and/or syllabus for any assigned activities.

OVERVIEW

You just can't stand it anymore. You've gained twenty-five pounds since you graduated from high school, and your daughter is getting married in six months! You know you need to lose the weight and with all the diet programs, pills, potions, and gadgets out there it will be a snap to lose those extra pounds in time for the wedding—or will it? How do you know the weight will stay off once you've lost it? What are the best weight-loss methods to use? Should you sign up for one of the commercial diet programs you see advertised on TV or go to a weight-loss clinic? What should you look for in a sound weight-loss program? These questions and more are answered in this lesson as it examines the treatment and health effects of weight control.

In addition to learning about good and poor treatments for obesity, you will also look at the causes and treatments for the eating disorders of anorexia nervosa and bulimia nervosa and gain some insight as to why certain people become victims of these disorders.

LEARNING OBJECTIVES

Upon completing this lesson, you should be able to:

1. Describe the causes of obesity.
2. Describe controversies in obesity treatment.
3. Explain poor treatment choices for weight loss.
4. Explain good treatment choices for weight loss.
5. Cite behavior modification techniques that would support a sound weight-loss program.
6. Describe strategies for gaining weight.
7. Explain ways to identify fad diets and other weight loss scams.
8. Compare fashion models to average American women in body size and/or body weight.
9. Explain how waist-to-hip ratios impact risk for health.
10. Explain the differences between anorexia nervosa and bulimia nervosa.

TEXT FOCUS POINTS

The following focus points are designed to help you get the most from your reading. Review them, then read the assignment. You may want to write notes to reinforce what you have learned.

1. Explain how each of the following can contribute to being overweight: fat cell development, fat cell metabolism, and set-point theory. Explain how genetics, overeating, and physical inactivity may cause obesity. Which contributors can people control?

2. What percent of the U.S. population is attempting to lose weight at any given time? State the prejudices against overweight people.

3. Explain how ineffective weight-loss treatment can add to a dieter's psychological burden. Describe how the following drugs function as weight-loss aids: phentermine, orlistat, St. John's wort, ephedrine, and herbal laxatives. Describe how herbal products are marketed for weight loss.

4. Under what condition is gastric surgery justified as an approach to weight loss? Describe immediate postsurgical complications associated with gastric surgery. What is liposuction?

5. What parameters are more useful than body weight in denoting successful weight loss? Based on initial weight, what do experts recommend as a reasonable percentage of weight loss? What is considered a reasonable time frame for a 10-percent loss of initial weight? What are the only considerations that need to be taken into account when designing a personal eating plan? What is the main characteristic of a weight-loss diet? How many kcalories per day should an adult expend through exercise and reduced food intake to support fat loss and retain lean tissue? What is the lower limit of kcalories per day an adult should consume to provide adequate nutrition when trying to lose weight? What is the goal with regard to food portions? What should be the center of meals and snacks? How long does it take for the satiety signal to reach the brain? What effect does a high-fat diet have on leptin levels? What effect does low kcalorie diets have on ghrelin levels? What effect does fat have on satiety? Why should people drink adequate water when trying to lose weight? List the tips for weight loss shown in table 9-3 entitled "Weight Loss Strategies." What is a reasonable rate of weight loss in a week for an overweight individual?

6. Explain why the best approach to weight management is a combination of physical activity and dieting. Describe the effect of activity on the following: BMR, body composition, appetite, psychological well-being, and spot reducing. What type of activity is most effective for weight management?

7. State behavior techniques or strategies that people can use to help with a weight-loss program. At what point in time do most people reach a plateau when losing weight? In order to maintain weight loss, what is the required number of kcalories per week that should be spent in physical activity?

8. What percent of the population in the United States is underweight? What are the risks associated with underweight? State strategies for weight gain.

9. What are fad diets? Compare the popular diets: Atkins, Zone, and South Beach. What are the side effects of low-carbohydrate diets? How can fad diets be identified?

VIDEO FOCUS POINTS

The following focus points are designed to help you get the most from the video segment of this lesson. Review them before watching the video. You may want to write notes to reinforce what you have learned.

1. How do average women in the United States compare in body weight and shape to fashion models?

2. What are the criteria for successful weight loss?

3. What are the health effects of high waist-to-hip ratios? How does having an "apple" shape impact risk for disease?

4. Describe the health effects of "yo-yo" dieting.

5. Define *gastric bypass*. What population are candidates for gastric bypass procedures?

6. Which eating disorder is most medically dangerous? What are the criteria for diagnosing and treating anorexia and bulimia? What are causes of eating disorders?

7. What is considered the most sound advice for successful weight loss? What advice should be given to women who have gained weight during pregnancy?

PROJECT

This project is required in order to complete the course successfully.

This portion of the lesson is designed to provide you with specific information about nutrient adequacy and balance, number of kcalories consumed, nutrient density, and variety of the foods you typically eat. After analyzing a three-to-seven-day food intake, you will be able to apply the information you have learned to make appropriate changes to improve eating habits.

Refer to the Diet Analysis Project for complete details and directions. Contact your instructor if you have any questions. The Diet Analysis Project is in the back of this student course guide.

Information provided in Chapters 2, 8, and 9 will be useful when answering questions pertaining to your Diet Analysis Project with regard to weight control.

Designing a Personal Weight Control Plan

NAME: ______________________________ ID #: ____________________

The purpose of this activity is to help you create an eating and exercise plan that will allow you to lose, gain, or maintain your healthiest weight.

Using your Diet Analysis Project and information in Chapter 9, follow the steps below to complete this activity:

1. Examine your *original* three-to-seven-day average printouts. The average should meet the following standards:
   A. The minimum recommended number of servings for each food group to ensure adequacy, balance, and nutrient density.
   B. A minimum of 1,200 kcalories to maintain metabolism and prevent nutrient deficiencies.
   C. Greater than 50 percent of total kcalories from carbohydrate to maintain glucose and spare protein.
   D. Less than 30 percent of total kcalories from fat to help reduce the risk for chronic diseases.
   E. Between 10 and 20 percent of total kcalories from protein for growth, fluid balance, enzyme production, electrolyte balance, etc.

2. If your original average did *not* meet the above criteria, then you need to make modifications to your individual days in one or more of the following ways:
   A. Add appropriate foods to each food group to reach the minimum number of servings.
   B. Adjust serving sizes up or down to increase or decrease kcalories.
   C. Use low-fat or nonfat foods to reduce total fat, saturated fat, and/or cholesterol intake.
   D. Substitute plant proteins for animal proteins to reduce fat and/or cholesterol intake (see the Related Activity in Lesson 8 for suggestions).

3. Print the following information from the diet analysis software program:
   A. The *original* three-day average reports
   B. The *modified* three-day average reports. Indicate on the printouts whether it is the original or the modified version. Please DO NOT print individual days—save a tree!

—Continued on the back

4. Answer the questions that follow (use additional paper if needed), then staple the questions and answers to the above printouts (from Step 3).

   A. Record your current weight and your desired weight, if different.
   ____________ Current weight ____________ Desired weight

   B. Select one of the following goals based on the weight that is healthiest for you:
   ______________ Weight maintenance
   ______________ Weight loss
   ______________ Weight gain

   C. ______________ How many kcalories per day will you consume to reach the above weight goal? (NOTE: When modifying your eating plan, stay within +/– 250 kcalories per day of the amount you indicate here.)

   D. Briefly explain what general modifications were made to your original three-to-seven-day food intake.

   E. List the foods you added (or deleted) and their serving sizes for each food group.

   Grains, cereals, breads:

   Fruits:

   Vegetables:

   Meat/meat alternates:

   Dairy:

5. Describe a detailed exercise plan, and include types of activity as well as frequency, duration, and intensity that would help achieve and/or maintain weight management. Cite page numbers in the text to support your answer.

Evaluating a Weight-Loss Program or Product

NAME: ______________________________ ID #: ________________________

This exercise will help you to evaluate a weight-loss program or product for safety and effectiveness.

DIRECTIONS:

A. Select any weight-loss product (or program), either over-the-counter (OTC) or by prescription. You may use the Internet to obtain information if necessary.

B. Answer the questions that follow. Use additional paper if needed.

1. What is the name of the product?

2. Is this product OTC or by prescription?

3. What is the cost of the product?

4. Is there a brochure, pamphlet, or any written instructions with the product?

5. What is the recommended intake or dosage?

6. List the first four ingredients and the amounts shown on the product label.

—Continued on the back

7. Is there anything unusual or unique about the product?

8. What is the product "claim" or "promise"?

9. What precautions, contraindications, or warnings are stated on the product?

10. List the exercise guidelines/recommendations that are presented with the product.

11. List the dietary guidelines/recommendations that are presented with the product.

12. If you (or someone you know) used this product for weight loss, briefly describe the experience. Include how much weight was lost, what the side effects were, whether the weight stayed off or was regained, etc.

PRACTICE TEST

The following items will help you check your understanding of this lesson. Compare your answers to the Answer Key at the end of the lesson. Review the course materials related to any incorrect answer.

Multiple Choice: Select the one choice that best answers the question.

1. All of the following describe the behavior of fat cells EXCEPT ________
   A. the number decreases when fat is lost from the body.
   B. the storage capacity for fat depends on both cell number and cell size.
   C. the number increases at a faster rate in obese children than in lean children.
   D. the number increases several-fold during the growth years and tapers off when reaching adult status.

2. Obese people have more ________ in their adipose cells.
   A. LPL
   B. HDL
   C. LDL
   D. EPI

3. Television watching contributes to obesity for all of the following reasons EXCEPT ________
   A. it promotes inactivity.
   B. it promotes between-meal snacking.
   C. it replaces time that could be spent eating.
   D. it gives high exposure to energy-dense foods featured in the commercial advertisements.

4. What is the recommended rate of weight loss per week for a 250 pound obese individual?
   A. One Pound
   B. Five Pounds
   C. 1 percent of Body Weight
   D. 5 percent of Body Weight

5. Which of the following would NOT be a recommendation of weight-reduction counseling?
   A. Reorganize some established behavior patterns.
   B. Use the exchange system to keep track of energy intake.
   C. Perform physical exercise to increase energy expenditure.
   D. Reduce daily energy intake to less than 1,000 kcalories to overcome the decrease in metabolic rate.

6. Physical activity is effective in weight management by providing all the following EXCEPT __________
   A. increases metabolism.
   B. increases lean muscle mass.
   C. increases stress.
   D. increases discretionary kcalorie allowance.

7. Compared to average weight for women in the United States, fashion models are __________
   A. 10 percent below the average.
   B. 18 percent below the average.
   C. 23 percent below the average.
   D. 35 percent below the average.

8. If people need to lose 50 pounds of weight and only achieve a 25-pound weight loss, the criteria established by the medical profession consider those people to be __________
   A. underachievers.
   B. failures.
   C. successful.
   D. hopeless.

9. People with high waist-to-hip ratios have a higher risk for heart disease and are characterized by shape as __________
   A. “apples.”
   B. “slugs.”
   C. “pears.”
   D. “sloths.”

10. A cycle of quick weight loss and weight regain which causes the body's metabolism to drop is known as __________
    A. cyclic regaining.
    B. dieting madness.
    C. yo-yo dieting.
    D. on-off cycling.

11. An extreme surgical procedure, such as a gastric bypass, produces dramatic weight loss in morbidly obese people by __________
    A. restricting the size of the stomach so it holds less food.
    B. stimulating the hypothalamus to not respond to excess food.
    C. shortening the GI tract so that food is not absorbed.
    D. reducing the length of the colon so food passes quickly.

12. Of the most common eating disorders, the one which is most medically dangerous is __________
    A. bulimia nervosa.
    B. anorexia nervosa.
    C. obesity nervosa.
    D. purging nervosa.

13. All of the following are acceptable treatments for bulimia EXCEPT __________
    A. helping the person normalize eating patterns.
    B. focusing on the person's social world.
    C. treating all people with bulimia in a hospital setting.
    D. changing the attitudes regarding weight and shape.

14. Experts believe that the most sound advice for losing weight is __________
    A. exercise more and eat a healthful diet.
    B. balance the kcalories in with the kcalories out.
    C. eat less than 1,000 kcalories per day.
    D. all of the above.

Matching: Select the letter next to the word or phrase that best matches the numbered statements. Answers are used only once.

_____15. An enzyme that promotes fat storage

_____16. A cosmetic surgical procedure

_____17. A fat cell hormone

_____18. Cause of an increase of basal metabolic rate

A. Leptin
B. Fasting
C. Liposuction
D. Gastric partitioning
E. Lipoprotein lipase
F. Exercise

Essay: Reflect on the item(s) listed below; one or more may be included on the exam.

19. List the major causes of obesity. Which ones can be controlled by dietary manipulations or behavior modification?

20. Present a sound diet plan for weight gain in the underweight person.

ANSWER KEY

The following provides the answers and references for the Practice Test questions.

Answer		Learning Objective	Reference
1.	A	LO 1	textbook, p. 262
2.	A	LO 1	textbook, p. 262
3.	C	LO 1	textbook, p. 267
4.	C	LO 5	textbook, pp. 272–273
5.	D	LO 3	textbook, pp. 274–275
6.	C	LO 4	textbook, pp. 277–279
7.	C	LO 8	Video
8.	C	LO 5	Video
9.	A	LO 9	Video
10.	C	LO 3	Video
11.	A	LO 2	Video
12.	B	LO 10	Video
13.	C	LO 10	Video
14.	A	LO 4	Video
15.	E	LO 1	textbook, p. 263
16.	C	LO 3	textbook, p. 271
17.	A	LO 1	textbook, p. 264
18.	F	LO 4	textbook, p. 278
19.		LO 1	textbook, pp. 262–267
20.		LO 6	textbook, pp. 285–286

Lesson 12

Vitamins: Water Soluble

LESSON ASSIGNMENTS

Text: Whitney and Rolfes, *Understanding Nutrition*, Chapter 10, "The Water-Soluble Vitamins: B Vitamins and Vitamin C," pp. 297–336

Video: "Vitamins: Water Soluble" from the series *Nutrition Pathways*

Project: "Diet Analysis Project"

Getting It Together:

The purpose of the "Getting It Together" section is to provide an outline of *new* or *difficult* material presented in the current lesson. This section provides practice in recalling lesson material and utilizes the textbook interactively. Use this section as a learning aid, to reinforce knowledge, or to organize information introduced in the lesson.

Related Activities:

These activities are not required unless your instructor assigns them. They are offered as suggestions to help you learn more about the material presented in this lesson. Refer to your syllabus to determine which of these activities have been assigned.
Complete the attached forms for the following related activities, and return them to your instructor according to the established deadline.

- ❒ "Calculating Niacin Intake"
- ❒ "Estimating Dietary Folate Equivalents"

Optional Web Activities:

Consult your instructor and/or syllabus for any assigned activities.

OVERVIEW

You have studied the energy-yielding nutrients—carbohydrate, fat, and protein. You examined the roles they play in the body, how they are digested, absorbed, and metabolized, and their effects on body weight. Now it is time to look at the non-energy-yielding nutrients—the vitamins and minerals. This lesson begins with vitamins B and C—the water-soluble vitamins—and examines their roles in the body, the deficiency and/or toxicity associated with their intake, and their major food sources. This lesson also explains how B vitamins work in concert to aid the body in energy production and how vitamin C works its antioxidant effects on the body.

LEARNING OBJECTIVES

Upon completing this lesson, you should be able to:

1. Explain the differences and similarities between vitamins and the energy-yielding nutrients.
2. Explain the primary function of the B vitamins and of their coenzymes.
3. Identify the eight B vitamins according to the following criteria:
    - The primary function(s) in the body
    - Coenzyme form(s) (if any)
    - Deficiency symptoms and/or disease (if any)
    - Toxicity symptoms and/or disease (if any)
    - Major food sources
    - Destructibility of the vitamin
4. Describe in general terms how B vitamins work together in the body.
5. Describe the primary function(s) of vitamin C, deficiency/toxicity symptoms and/or disease, and major food sources.
6. Explain the antioxidant effects of vitamin C on the body.
7. Explain how vitamin losses can be prevented during handling, storage, and preparation.

TEXT FOCUS POINTS

The following focus points are designed to help you get the most from your reading. Review them, then read the assignment. You may want to write notes to reinforce what you have learned.

1. How do vitamins differ from energy-yielding nutrients with regard to structure, function, and amounts found in foods? What two factors affect the availability of vitamins from foods? Define *bioavailability*. Upon what does bioavailability depend? Why should care be taken when storing or cooking foods that contain water-soluble vitamins? How does solubility affect vitamin absorption, transport, and storage? How and why did the Committee on Dietary Reference Intakes address high doses of vitamins?

2. How do the B vitamins function in general in the body? How do B vitamins act as coenzymes?

3. What is the coenzyme form of thiamin? What is the primary function of the coenzyme? What disease is associated with thiamin deficiency, and how was it discovered? What symptoms are associated with thiamin deficiency disease? What foods are exceptional sources of thiamin?

4. What are the two coenzyme forms of riboflavin? How do riboflavin's coenzymes function in the body? What systems in the body are affected by riboflavin deficiency? What food group provides the greatest contribution of riboflavin to the diet? How do light and heat affect riboflavin?

5. What are the two coenzyme forms of niacin? What role does niacin play in metabolism? Name the amino acid that can be used in the body to make niacin. Define *niacin equivalents*. What disease is associated with niacin deficiency? What type of diet will produce niacin deficiency? What are "the four Ds" associated with niacin deficiency? What is *niacin flush*? What disease or medical condition is sometimes treated with niacin? What foods are the best sources of preformed niacin? What types of foods are best for niacin equivalents in the diet?

6. How does biotin generally function in energy production? How can a biotin deficiency be induced in animals and humans, and what are the effects? What are the sources of biotin for humans?

7. What is the coenzyme form of pantothenic acid? Generally speaking, how does the coenzyme function? What are particularly good food sources of pantothenic acid?

8. What is the coenzyme form and function of vitamin B_6? How does B_6 affect amino acid metabolism? Where is B_6 stored in the body? How does alcohol intake affect B_6 status? How does the medication isoniazid affect B_6 status? What are the symptoms of B_6 deficiency? How did researchers discover that B_6 had toxic levels? What are the toxic effects of B_6? What are the best food sources for B_6?

9. What is the coenzyme form of folate? What is the main function of folate's coenzyme? With what other B vitamin does folate interact? How does the body handle excess folate? What is the effect of alcohol abuse on folate status? Define *dietary folate equivalents*. Why are folate recommendations increased for pregnant women? Explain the importance of folate in reducing the risks of neural tube defects. What is the recommendation for folate supplementation prior to and during pregnancy? What connection exists between folate and heart disease? What are the two first symptoms of folate deficiency? Describe how anticancer drugs, aspirin, antacids, oral contraceptives, and smoking interact with folate. What foods have abundant amounts of folate?

10. Describe the relationship between B_{12} and folate. What is *intrinsic factor* and how does it affect B_{12} absorption? What is the most likely reason for developing a B_{12} deficiency? Define *atrophic gastritis*. What is the condition caused by B_{12} deficiency? What other B vitamin deficiency results from a B_{12} deficiency? In what foods is B_{12} found almost exclusively? How does B_{12} intake possibly impact vegan vegetarians?

11. Describe the roles of B vitamin coenzymes in metabolism. What areas of the body seem to be especially sensitive to vitamin B deficiencies? Under what circumstances do toxicities of the B vitamins occur? In what food groups do the B vitamins appear?

12. What is the disease caused by vitamin C deficiency? Define *antiscorbutic factor* and *antioxidant*. How does vitamin C act as an antioxidant? How does vitamin C function with regard to iron? How does vitamin C function with regard to collagen formation? In what compounds does vitamin C serve as a cofactor in their synthesis? What conditions or stressors are known to increase

the need for vitamin C? What is the RDA for vitamin C for men and women? Cite the recommendations for vitamin C for each of the following: scurvy, smokers, and people who have undergone surgery or extensive burns. Why is there an increased RDA for people who smoke or are recovering from surgery or burns? What are the earliest signs and symptoms of vitamin C deficiency? What are the symptoms associated with vitamin C toxicity? Cite several instances in which large amounts of vitamin C can produce toxic symptoms. What is the *upper level* of vitamin C intake for adults (see Figure 10-17)? What are the foods that provide generous amounts of vitamin C?

VIDEO FOCUS POINTS

The following focus points are designed to help you get the most from the video segment of this lesson. Review them before watching the video. You may want to write notes to reinforce what you have learned.

1. How do B vitamins interact with energy-yielding foods, especially carbohydrate and fat?

2. How might vitamin C affect the immune system?

3. What are suggestions to prevent vitamin loss during handling, storage, and preparation of vegetables for home use and for extended periods?

PROJECT

This project is required in order to complete the course successfully.

This portion of the lesson is designed to provide you with specific information about nutrient adequacy and balance, number of kcalories consumed, nutrient density, and variety of the foods you typically eat. After analyzing a three-to-seven-day food intake, you will be able to apply the information you have learned to make appropriate changes to improve eating habits.

Refer to the Diet Analysis Project for complete details and directions. Contact your instructor if you have any questions. The Diet Analysis Project is in the back of this student course guide.

Information provided in Chapter 10 will be useful when answering questions pertaining to your Diet Analysis Project with regard to water-soluble vitamins.

GETTING IT TOGETHER

Fill in the Table of Water-Soluble Vitamins

These tables may be used as study aids, submitted as part of course requirements, or used as extra credit opportunities as determined by your instructor. You may wish to enlarge the tables before completing them.

DIRECTIONS: In the space provided, write in two to four of each of the following: the major functions, deficiency disease and/or symptoms, toxicity symptoms, and the major food sources *found in your diet* for each of the B vitamins and vitamin C. Refer to your Diet Analysis Project for your food sources. If a characteristic does not apply, write NA in the space.

For vitamins not analyzed in your Diet Analysis Project, list appropriate food sources.

Vitamin (Coenzyme)	Function(s) of Vitamin	Deficiency Disease and/or Symptoms	Toxicity Symptoms	Major Food Sources in Your Diet
B_1, Thiamin/ (TPP)	1. 2. 3. 4.	1. 2. 3. 4.	1. 2. 3. 4.	1. 2. 3. 4.
B_2, Riboflavin/ (FMN, FAD)	1. 2. 3. 4.	1. 2. 3. 4.	1. 2. 3. 4.	1. 2. 3. 4.
B_3 Niacin/ (NAD, NADP)	1. 2. 3. 4.	1. 2. 3. 4.	1. 2. 3. 4.	1. 2. 3. 4.

—Continued on the back

GETTING IT TOGETHER—*Continued*

Vitamin (Coenzyme)	Function(s) of Vitamin	Deficiency Disease and/or Symptoms	Toxicity Symptoms	Major Food Sources in Your Diet
Biotin	1. 2. 3. 4.	1. 2. 3. 4.	1. 2. 3. 4.	1. 2. 3. 4.
Pantothenic Acid/ (CoA)	1. 2. 3. 4.	1. 2. 3. 4.	1. 2. 3. 4.	1. 2. 3. 4.
B_6, Pyridoxal, -xine, -xamine/ (PLP)	1. 2. 3. 4.	1. 2. 3. 4.	1. 2. 3. 4.	1. 2. 3. 4.
Folate, Folic Acid, Folacin, PGA/(THF)	1. 2. 3. 4.	1. 2. 3. 4.	1. 2. 3. 4.	1. 2. 3. 4.
B_{12}, Cobalamin	1. 2. 3. 4.	1. 2. 3. 4.	1. 2. 3. 4.	1. 2. 3. 4.
C, Ascorbic Acid	1. 2. 3. 4.	1. 2. 3. 4.	1. 2. 3. 4.	1. 2. 3. 4.

Calculating Niacin Intake

NAME: ________________________________ ID #: ________________________

We get niacin from foods rich in that B vitamin and from niacin equivalents (niacin derived from the amino acid tryptophan). The purpose of this activity is to demonstrate how many milligrams of niacin are in the foods consumed in a typical day.

Use the diet analysis software or refer to Appendix H to complete this activity. Refer to Chapter 10 in your textbook to answer questions.

DIRECTIONS:

1. Calculate how many grams of protein and milligrams of niacin are found in each of the following foods.
2. Answer the questions that follow.

BREAKFAST	**G Protein**	**Mg Niacin**
1 cup oatmeal		
1 cup skim milk		
1 small apple		
1 tsp sugar		
LUNCH		
2 oz lean roast beef		
2 slices white bread		
1 tbs mustard		
1 oz potato chips		
DINNER		
4 oz chicken breast		
½ cup green beans		
1 cup brown rice		
1 cup low-fat ice cream		
TOTALS		

—Continued on the back

Calculating Niacin Intake—*Continued*

1. __________ What is your RDA for protein according to the RDA Table?

2. __________ How many total grams of protein are in the list of foods?

3. __________ Determine milligrams of tryptophan found in protein as follows:
   - Subtract your RDA for protein (Step 1) from the total grams of dietary protein (Step 2). This is called "leftover" protein.
   - Divide by 100 to obtain grams of tryptophan.
   - Multiply by 1,000 to obtain milligrams of tryptophan. (*Show your calculations below.*)

4. __________ Determine niacin equivalents as follows: Divide milligrams of tryptophan (Step 3) by 60. (*Show your calculations below.*)

5. __________ What is your RDA for niacin according to the RDA Table?

6. __________ How many total milligrams of niacin (preformed niacin) are in the list of foods?

7. __________ Determine total niacin intake as follows: Add the amount of preformed niacin (Step 6) to niacin equivalents (Step 4). (*Show your calculations below.*)

Cite page numbers in the text to support your answers to the following questions:

8. Identify the niacin deficiency disease and name the "four Ds" associated with it.

9. Based upon the information obtained from this activity and assuming you consumed similar amounts of niacin daily, explain whether the total niacin intake (Step 7) would provide you with enough niacin to prevent niacin deficiency symptoms.

Estimating Dietary Folate Equivalents

NAME: ______________________________ ID #: ______________________

The purpose of this activity is to determine whether folate from synthetic sources along with naturally occurring folate supplies enough folate for pregnant women.

Refer to Chapter 10 in your textbook to help answer the following questions.

PROBLEM:
A pregnant woman takes a supplement containing 100 micrograms (μg) of synthetic folate. She obtains more synthetic folate from one cup of fortified cereal (100 μg), one slice of fortified bread (20 μg), and one cup of fortified pasta (60 μg). In addition to the synthetic folate, she also consumes 80 μg of natural folate from other foods.

1. __________ What is the total amount of *synthetic* folate consumed by the pregnant woman? (*Show calculations below.*)

2. __________ Calculate the dietary folate equivalents (DFE) by completing the formula below: (NOTE: Folate is expressed in terms of DFE because synthetic folate from supplemental folate and fortified foods is absorbed at 1.7 times the rate of naturally occurring folate.)

   ______ μg synthetic folate x 1.7 = _______ μg DFE

3. __________ What is the total amount of *natural* folate consumed by the pregnant woman?

4. __________ Calculate the total folate consumed by the pregnant woman by completing the formula below:

   _______ μg DFE + _______ μg natural folate = _________ μg total folate

—*Continued on the back*

5. ____________ What is the RDA for folate for pregnant women? (Refer to the RDA Table in the textbook.)

6. Explain whether the pregnant woman in this example obtained enough folate to support pregnancy. (Cite page numbers in the text to support your answer.)

PRACTICE TEST

The following items will help you check your understanding of this lesson. Compare your answers to the Answer Key at the end of the lesson. Review the course materials related to any incorrect answer.

Multiple Choice: Select the one choice that best answers the question.

1. What is the primary excretory route for the water-soluble vitamins?
   A. Bile
   B. Kidney
   C. Intestine
   D. Perspiration

2. What is the primary function of the B vitamins?
   A. Energy source
   B. Anticoagulation
   C. Coenzyme participation
   D. Antibody stabilization

3. Which of the following diets is most likely to lead to beriberi?
   A. Low intakes of whole grains
   B. High intakes of polished rice
   C. High intakes of unrefined rice
   D. Low intakes of enriched grains

4. Of the following, which is the richest food source of thiamin?
   A. Milk
   B. Pork
   C. Lettuce
   D. Refined rice

5. When the diet contains an adequate amount of protein, what amino acid can be used by the body to synthesize niacin?
   A. Lysine
   B. Valine
   C. Tryptophan
   D. Phenylalanine

6. Which of the following is a feature of niacin nutrition?
   A. Low doses may lead to kidney stones.
   B. High doses may lower blood cholesterol.
   C. Low doses may lead to heartburn and low blood pressure.
   D. High doses may elevate red blood cell count in mildly anemic individuals.

7. Which of the following activities is shared by vitamin B_{12} and folate?
   A. Both are required for nucleic acid synthesis.
   B. Both require intrinsic factor for their release from food proteins.
   C. Both are found in significant amounts in green leafy vegetables.
   D. Both are considered problem nutrients for strict vegetarians.

8. Which of the following is a type of neural tube defect?
   A. Scurvy
   B. Beriberi
   C. Pellagra
   D. Spina bifida

9. Which of the following represents the results of well-controlled studies of vitamin C supplementation on the resistance to, and recovery from, colds?
   A. There was a reduction in the duration of colds by 50 percent on the average.
   B. There was only a minor effect on reducing the number and severity of colds.
   C. There was a significant reduction in the duration of colds only in people who consumed at least one gram a day.
   D. There was a significant reduction in the number of colds only in people who consumed more than three grams per day.

10. Which of the following vitamins is known to deactivate histamine, a substance that causes nasal congestion?
   A. Niacin
   B. Vitamin E
   C. Vitamin C
   D. Vitamin B_{12}

11. In which of the following ways do B vitamins interact with energy-yielding foods?
    A. They provide an essential source of additional energy.
    B. They are not a source of energy, but they are needed to convert the foods we eat into energy.
    C. B vitamins only interact with fatty acids to produce energy.
    D. B vitamins only interact with carbohydrates to produce energy.

12. As far as preparing vegetables is concerned, the key in preventing vitamin losses is to __________
    A. cook until soft to the touch.
    B. not overcook—keep slightly crisp.
    C. use a lot of water when cooking.
    D. drink the water in which the vegetables were cooked.

Fill in the Blank: Insert the correct word or words in the blank for each item.

13. When studying allergies and aging, research has found a positive effect from vitamin ____.

Matching: Select the letter next to the word or phrase that best matches the numbered statements. Answers are used only once.

______14. Used for synthesis of niacin

______15. Toxicity from this vitamin is known to cause irreversible nerve damage

______16. One of the first symptoms of folate deficiency

______17. Vitamin C is required for the synthesis of this substance

A. Anemia
B. Collagen
C. Tryptophan
D. Vitamin B_6
E. Vitamin B_{12}

Essay: Reflect on the item(s) listed below; one or more may be included on the exam.

18. Under what circumstances can water-soluble vitamins be toxic? Cite several examples.

19. Discuss the roles of the B vitamins in energy metabolism.

20. Under what conditions and for what reasons would intakes of vitamin C above the RDA be desirable?

ANSWER KEY

The following provides the answers and references for the Practice Test questions.

	Answer	Learning Objective	Reference
1.	B	LO 1	textbook, p. 299
2.	C	LO 2	textbook, pp. 300–301
3.	B	LO 3	textbook, pp. 301–302
4.	B	LO 3	textbook, p. 303
5.	C	LO 3	textbook, p. 305
6.	B	LO 3	textbook, pp. 306–307
7.	A	LO 3	textbook, p. 311
8.	D	LO 3	textbook, p. 313
9.	B	LO 6	textbook, p. 324
10.	C	LO 6	textbook, p. 324
11.	B	LO 4	Video
12.	B	LO 7	Video
13.	C	LO 5	Video
14.	C	LO 3	textbook, p. 305
15.	D	LO 3	textbook, p. 310
16.	A	LO 3	textbook, p. 314
17.	B	LO 5	textbook, p. 323
18.		LO 3	textbook, pp. 299–300
19.		LO 4	textbook, pp. 300–301, 319
20.		LO 5	textbook, p. 324

Lesson 13

Vitamins: Fat Soluble

LESSON ASSIGNMENTS

Text: Whitney and Rolfes, *Understanding Nutrition*, Chapter 11, "The Fat-Soluble Vitamins: A, D, E, and K," pp. 339–363

Video: "Vitamins: Fat Soluble" from the series *Nutrition Pathways*

Project: "Diet Analysis Project"

Getting It Together:

The purpose of the "Getting It Together" section is to provide an outline of *new* or *difficult* material presented in the current lesson. This section provides practice in recalling lesson material and utilizes the textbook interactively. Use this section as a learning aid, to reinforce knowledge, or to organize information introduced in the lesson.

Optional Web Activities:

Consult your instructor and/or syllabus for any assigned activities.

OVERVIEW

In this lesson, you will learn that there are many differences between the fat-soluble and water-soluble vitamins. Fat-soluble vitamins are found in the oils and fats of foods, are insoluble in water, need bile to be digested, require chylomicrons for transport, go through the lymphatic system before entering the bloodstream, and are stored in the liver and fat tissues. You will learn what happens when you don't have enough of the fat-soluble vitamins in your diet. Because they are not readily excreted, you will also see how taking excessive fat-soluble vitamins can have toxic effects and why we only need to take them over a period of several days rather than daily. Finally, you will recognize the function of vitamin E as an antioxidant.

LEARNING OBJECTIVES

Upon completing this lesson, you should be able to:

1. Explain the differences between water- and fat-soluble vitamins.

2. Identify the fat-soluble vitamins according to the following criteria:
   - The primary function(s) in the body
   - Different forms (if any) and their function(s)
   - Deficiency symptoms and/or disease (if any)
   - Toxicity symptoms and/or disease (if any)
   - Major food sources

3. Explain how vitamin E acts as an antioxidant.

4. State the pros and cons of antioxidant supplementation.

TEXT FOCUS POINTS

The following focus points are designed to help you get the most from your reading. Review them, then read the assignment. You may want to write notes to reinforce what you have learned.

1. Describe how fat-soluble vitamins differ from water-soluble vitamins for the following: solubility, absorption/transport, storage, excretion, frequency of intake, and risk for toxicity.

2. Define *beta-carotene*. Name the three active forms of vitamin A in the body. What is the collective term for the compounds of vitamin A? What is the transport protein that carries vitamin A to the blood from the liver?

3. What are the three major roles of vitamin A in the body? Explain how each of the following forms of vitamin A functions: retinol, retinal, and retinoic acid. Describe the two roles of vitamin A in the eyes. Define *rhodopsin* and *opsin*. How are retinal and opsin affected by light? How does visual activity affect retinal use? How much of the body's vitamin A is in the retina? In what function does most of the vitamin A participate? How does vitamin A function in reproduction and bone remodeling? In what capacity does some beta-carotene act? How does beta-carotene act as an antioxidant?

4. Where is 90 percent of vitamin A stored in the body? What population is most susceptible to vitamin A deficiency? What is the most common cause of death in someone with a vitamin A deficiency? What is the first detectable sign of vitamin A deficiency? What is night blindness? Define the following terms: *xerophthalmia*, *xerosis*, *keratomalacia*, *keratin*, and *keratinization*.

5. What causes vitamin A toxicity? What can happen if a person eats too many foods containing beta-carotene? What is the recommendation for limiting vitamin A supplements for pregnant women and why? What is the recommendation regarding vitamin A and acne treatments? What foods are the best sources of preformed vitamin A? What foods contain carotenoids? What carotenoid has the highest vitamin A activity? What color in vegetables is typically associated with Vitamin A content? Why is caution advised for people who frequently consume large amounts of liver?

6. Why is vitamin D not considered an essential nutrient? How is the active form of vitamin D produced in the body? What nutrient is the precursor for vitamin D? What is the active form of vitamin D? What organs are the primary targets of vitamin D activity? What is the special function of vitamin D in bone growth?

7. What happens in the intestinal cells with a vitamin D deficiency? What is rickets? What is osteomalacia? Why is vitamin D deficiency more likely to occur in elderly people? How does vitamin D compare to other fat-soluble vitamins with regard to toxic effects? What are the symptoms of vitamin D toxicity? What are the best food sources of vitamin D? What populations would most likely need supplementation of vitamin D? How much time in the sun is needed for most people to maintain vitamin D status? How much time in the sun is needed for dark-skinned versus light-skinned people for vitamin D synthesis to occur?

8. Explain the role of vitamin E as an antioxidant. What is the classic sign of vitamin E deficiency? How does vitamin E toxicity compare to toxicity associated with other fat-soluble vitamins? What can happen with extremely high doses of vitamin E? What is the most biologically active vitamin E compound? What are the best food sources of vitamin E? Why do most processed and convenience foods lack vitamin E?

9. What nonfood source contributes to vitamin K that the body can absorb? What is the primary function of vitamin K? Describe how vitamin K

functions in blood clot formation. What is hemorrhagic disease? How does vitamin K function in bone formation? Under what two circumstances can vitamin K deficiency develop? Why are newborns given vitamin K at birth? When can vitamin K toxicity occur? What are the best food sources of vitamin K? How much vitamin K can be synthesized by GI tract bacteria?

VIDEO FOCUS POINTS

The following focus points are designed to help you get the most from the video segment of this lesson. Review them before watching the video. You may want to write notes to reinforce what you have learned.

1. How can vitamin A toxicity affect the brain?

2. Provide examples of oxidation reactions.

3. Under what circumstances might a doctor prescribe antioxidant supplementation? What is vitamin E's chief role as an antioxidant? What is lipid peroxidation? What diseases might be prevented through the use of vitamins E and C?

4. What is the advice regarding antioxidant supplementation, exercise, and immunity?

PROJECT

This project is required in order to complete the course successfully.

This portion of the lesson is designed to provide you with specific information about nutrient adequacy and balance, number of kcalories consumed, nutrient density, and variety of the foods you typically eat. After analyzing a three-to-seven-day food intake, you will be able to apply the information you have learned to make appropriate changes to improve eating habits.

Refer to the Diet Analysis Project for complete details and directions. Contact your instructor if you have any questions. The Diet Analysis Project is in the back of this student course guide.

Information provided in Chapter 11 will be useful when answering questions pertaining to your Diet Analysis Project with regard to fat-soluble vitamin intake.

GETTING IT TOGETHER

Fill in the Table of Fat-Soluble Vitamins

These tables may be used as study aids, submitted as part of course requirements, or used as extra credit opportunities as determined by your instructor. You may wish to enlarge the tables before completing them.

DIRECTIONS: In the space provided, write in two to four of each of the following: the major functions, deficiency disease and/or symptoms, toxicity symptoms, and the major food sources *found in your diet* for each of the fat-soluble vitamins. Refer to your Diet Analysis Project for your food sources. If a characteristic does not apply, write NA in the space.

NOTE: Vitamin A forms are treated separately.

Name of Vitamin	Function(s) of Vitamin	Deficiency Disease and/or Symptoms	Toxicity Symptoms	Major Food Sources in your Diet
A: Retinol	1. 2. 3. 4.	1. 2. 3. 4.	1. 2. 3. 4.	1. 2. 3. 4.
A: Retinal	1. 2. 3. 4.	1. 2. 3. 4.	1. 2. 3. 4.	1. 2. 3. 4.
A: Retinoic Acid	1. 2. 3. 4.	1. 2. 3. 4.	1. 2. 3. 4.	1. 2. 3. 4.

—*Continued on next page*

GETTING IT TOGETHER—*Continued*

Name of Vitamin	Function(s) of Vitamin	Deficiency Disease and/or Symptoms	Toxicity Symptoms	Major Food Sources in Your Diet
A: Beta-carotene	1. 2. 3. 4.	1. 2. 3. 4.	1. 2. 3. 4.	1. 2. 3. 4.
D: (1,25-Dihydroxy-cholecalciferol; Calcitriol; Vitamin D_3)	1. 2. 3. 4.	1. 2. 3. 4.	1. 2. 3. 4.	1. 2. 3. 4.
E: (Alpha-tocopherol)	1. 2. 3. 4.	1. 2. 3. 4.	1. 2. 3. 4.	1. 2. 3. 4.
K	1. 2. 3. 4.	1. 2. 3. 4.	1. 2. 3. 4.	1. 2. 3. 4.

PRACTICE TEST

The following items will help you check your understanding of this lesson. Compare your answers to the Answer Key at the end of the lesson. Review the course materials related to any incorrect answer.

Multiple Choice: Select the one choice that best answers the question.

1. Which of the following is NOT among the features of the fat-soluble vitamins?
   A. Require bile for absorption
   B. Found in the fat and oily parts of foods
   C. Transported permanently to the liver and adipose tissue
   D. Pose a greater risk for developing a toxicity than water-soluble vitamins

2. Which of the following functions of vitamin A accounts for most of the body's need for the vitamin?
   A. Promoting good night vision
   B. Assisting in immune reactions
   C. Promoting the growth of bones
   D. Maintaining mucous membranes

3. Vitamin A supplements are helpful in treating which of the following conditions?
   A. Acne
   B. Rickets
   C. Osteomalacia
   D. Night blindness

4. In what system would the effects of a vitamin D deficiency be most readily observed?
   A. Nervous
   B. Skeletal
   C. Muscular
   D. Circulatory

5. The main function of vitamin E in the body is to act as a(n) ________
   A. coenzyme.
   B. peroxide.
   C. antioxidant.
   D. free radical.

6. Among the following, which contains the highest concentration of vitamin E?
   A. Butter
   B. Carrots
   C. Milk fat
   D. Corn oil

7. What vitamin is routinely given as a single dose to newborns?
   A. Vitamin A
   B. Vitamin E
   C. Vitamin K
   D. Vitamin B_{12}

8. What is a free radical?
   A. A highly reactive, unstable molecule that contains unpaired electrons
   B. An antioxidant substance that prevents accumulation of cell-damaging oxides
   C. A substance in food that interacts with nutrients to decrease their utilization
   D. A nutrient in excess of body needs that the body is free to degrade with no consequence

9. If taken in excess, vitamin A can produce adverse effects, such as __________
   A. intracranial pressure in the brain which can cause headaches.
   B. blurred vision which can cause headaches.
   C. bone mineralization which can cause rickets.
   D. all of the above.

10. Rust forming on a metal surface or butter turning rancid are examples of ________
    A. an oxidation reaction which produces undesirable results.
    B. a mineral interacting with a hard surface.
    C. iron particles interacting with water molecules.
    D. B and C.

11. Several conditions and/or diseases such as some cancers, aging, and heart disease might be prevented in the future through the use of __________
    A. antioxidant supplementation with vitamins C and E.
    B. medications that have yet to be discovered.
    C. new exercise equipment that is currently being developed.
    D. all of the above.

12. Young patients who have heart disease who do not have risk factors that can be treated might be advised to take __________
    A. vitamin A supplements.
    B. vitamin E supplements.
    C. vitamin D supplements.
    D. multiple vitamins.

13. Vitamin E supplements are often taken by runners to help prevent damage to lungs caused by __________
    A. heavy exercise during cold weather.
    B. exercising when air pollution is high.
    C. exercising when they have a respiratory infection.
    D. all of the above.

Fill in the Blank: Insert the correct word or words in the blank for each item.

14. When rust forms on an iron railing or cooking oil smells rancid, the undesirable effect is caused by a process known as ____________________.

15. The body system that appears to improve when elderly people take vitamin E and other antioxidants is __________________________.

Matching: Select the letter next to the word or phrase that best matches the numbered statements. Answers are used only once.

______ 16. Transport protein of vitamin A

______ 17. Pigment in carrots and pumpkins

______ 18. Another term for blood clotting

______ 19. Food source of vitamin K

A. Carotene
B. Coagulation
C. Green leafy vegetables
D. Retinol-binding protein
E. Keratin

Essay: Reflect on the item(s) listed below; one or more may be included on the exam.

20. Describe how the body can synthesize active vitamin D with the help of sunlight.

21. Describe the known functions of vitamin E and false claims of vitamin E supplementation.

22. Discuss the pros and cons of taking antioxidant supplements.

ANSWER KEY

The following provides the answers and references for the Practice Test questions.

	Answer	Learning Objective	Reference
1.	C	LO 1	p. 339
2.	D	LO 2	p. 342
3.	D	LO 2	p. 343
4.	B	LO 2	pp. 348–349
5.	C	LO 3	p. 353
6.	D	LO 2	p. 354
7.	C	LO 2	p. 355
8.	A	LO 3	p. 353
9.	A	LO 2	Video
10.	A	LO 3	Video
11.	A	LO 4	Video
12.	B	LO 3	Video
13.	B	LO 4	Video
14.	oxidation	LO 3	Video
15.	the immune system (immunity, immune response)	LO 3	Video
16.	D	LO 2	p. 340
17.	A	LO 2	p. 340
18.	B	LO 2	p. 354
19.	C	LO 2	p. 356
20.		LO 2	pp. 350–351
21.		LO 4	pp. 353–354
22.		LO 4	pp. 360–363

Lesson 14

Major Minerals and Water

LESSON ASSIGNMENTS

Text: Whitney and Rolfes, *Understanding Nutrition*, Chapter 12, "Water and the Major Minerals," pp. 367–394

Video: "Major Minerals and Water" from the series *Nutrition Pathways*

Project: "Diet Analysis Project"

Getting It Together:
The purpose of the "Getting It Together" section is to provide an outline of *new* or *difficult* material presented in the current lesson. This section provides practice in recalling lesson material and utilizes the textbook interactively. Use this section as a learning aid, to reinforce knowledge, or to organize information introduced in the lesson.

Optional Web Activities:
Consult your instructor and/or syllabus for any assigned activities.

OVERVIEW

"Drink eight glasses of water a day." You've heard this admonition before, so you know that water must be good for you. You could literally live for weeks without food, but you would not live long, maybe a few days, without water—it's that important to life. In this lesson, you will realize how critical water is to normal, healthy functioning. In addition, you will examine other nutrients that are necessary for good health—the major minerals. You will discover that major minerals are differentiated from vitamins not only in amounts needed for bodily functions but also in their chemical structure: minerals are inorganic compounds that appear in our bodies in the same form as they appear in nature—indestructible—and essential to life. They have similar qualities to vitamins in that some need assistance in being absorbed and transported to cells, whereas others do not, and some minerals, like some vitamins, can be toxic if taken in excess of needs. Further, you will also learn what deficiencies can develop when both minerals and water are not consumed in adequate amounts. Finally, you will focus attention on the development, symptoms, and treatment of osteoporosis, a condition directly related to the major mineral calcium.

LEARNING OBJECTIVES

Upon completing this lesson, you should be able to:

1. Explain the function of fluids/water in the body.
2. Describe how water is balanced in the body and what the recommendations are for water intake.
3. Briefly describe how blood volume and pressure are regulated.
4. Briefly explain how fluids and electrolytes are regulated, including acid-base balance.
5. Identify the major minerals according to the following criteria:
    - Chief functions in the body
    - Deficiency symptoms and/or disease
    - Toxicity symptoms and/or disease
    - Major food sources
6. Cite the recommended minimum and maximum intake of sodium.
7. Describe the consequences of dehydration in the body.
8. Describe the development, symptoms, and treatment of osteoporosis.

TEXT FOCUS POINTS

The following focus points are designed to help you get the most from your reading. Review them, then read the assignment. You may want to write notes to reinforce what you have learned.

1. How does intracellular fluid differ from interstitial fluid and extracellular fluid? Cite the functions of water in body fluids.

2. How much of an adult's body consists of water? What factors influence water intake? Explain how the following body systems sense water needs: mouth, hypothalamus, and nerves. What are the sources of water for the body? How much water per day does a person derive from these sources? How much water per day must the body excrete to rid itself of waste products? How much water per day does a person need under normal environmental circumstances? How does adequate water intake impact health? Does the type of water make a difference?

3. How does each of the following function in regulating blood volume and/or blood pressure: antidiuretic hormone, renin, angiotensin, and aldosterone?

4. Define the following terms: *electrolyte*, *electrolyte solution*, *osmotic pressure*, and *sodium-potassium pump*. What two sites regulate fluid/electrolyte balance? What are circumstances which could cause electrolyte/fluid imbalances? Which minerals are the first to be lost through fluid losses? Under what three circumstances would medical intervention be required to replace lost fluids?

5. What are buffers? How does carbon dioxide function in acid-base balance? What organ plays the primary role in acid-base balance?

6. Cite the differences between minerals and vitamins in the following ways: absorption, transport, and toxicity.

7. How does sodium function in the body? What is the recommended minimum sodium intake per day? What is the health recommendation regarding daily salt intake? What is the connection between sodium and hypertension? What is the connection between sodium and osteoporosis? Which foods provide the most sodium to the diet? Describe how a sodium deficiency might develop. What are the symptoms of sodium toxicity?

8. What are the functions of chloride in the body? What is chloride usually associated with in the diet? How may chloride be lost from the body?

9. What are the roles of potassium in the body? What foods provide abundant amounts of potassium? What role does potassium play in hypertension? What conditions can produce a potassium deficiency? What is an early sign of potassium deficiency? What are the toxic effects of potassium?

10. How much calcium is found in the body, and where is it found? What two roles does calcium play in the bones? What is hydroxyapatite? How does the calcium ion function in body fluids? What role does calcium play in hypertension? What body systems help raise blood calcium when it falls too low? How do these body systems help achieve calcium balance? What hormones are involved in regulation of blood calcium? What factors help increase calcium absorption? What are the recommendations for calcium intake? What food provides the single most abundant source of calcium? What are other food sources of available calcium? What deficiency develops with inadequate calcium intake?

11. How much phosphorus is found in the body, and where is it found? Describe the roles phosphorus plays in the body. What are the best sources of phosphorus in the diet? How does phosphorus interact with calcium?

12. How much magnesium is found in the body, and where is it found? How does magnesium function in the body? What foods provide the most magnesium to the diet? What conditions might produce magnesium deficiency? What role does magnesium play in hypertension?

13. What role does sulfur play in the body? What foods contain sulfur? How can sulfur deficiency develop?

VIDEO FOCUS POINTS

The following focus points are designed to help you get the most from the video segment of this lesson. Review them before watching the video. You may want to write notes to reinforce what you have learned.

1. How much body weight in the form of water could be lost during an hour's worth of exercise? What percent of water loss could result in a life-threatening situation? What is the very first sign of dehydration? What physical symptoms might be displayed by a child who is dehydrated? What advice is given to parents of a dehydrated child?

2. What amount of sodium per day is needed for an average adult?

3. How is osteoporosis characterized? Which population is most affected? What are the most common causes of osteoporosis? What are the risk factors associated with osteoporosis? What is the treatment for osteoporosis with regard to the following: exercise, calcium, salt, and animal products? What is a recent, novel approach to treatment, and what have been the results of the treatment? How can osteoporosis be prevented? What is often the first symptom of osteoporosis?

4. How can increased water intake impact sodium's adverse effects on the body?

PROJECT

This project is required in order to complete the course successfully.

This portion of the lesson is designed to provide you with specific information about nutrient adequacy and balance, number of kcalories consumed, nutrient density, and variety of the foods you typically eat. After analyzing a three-to-seven-day food intake, you will be able to apply the information you have learned to make appropriate changes to improve eating habits.

Refer to the Diet Analysis Project for complete details and directions. Contact your instructor if you have any questions. The Diet Analysis Project is in the back of this student course guide.

Information provided in Chapter 12 will be useful when answering questions pertaining to your Diet Analysis Project with regard to major minerals and water.

GETTING IT TOGETHER

Fill in the Table for Water

These tables may be used as study aids, submitted as part of course requirements, or used as extra credit opportunities as determined by your instructor. You may wish to enlarge the tables before completing them.

In the space provided, list two to four of the following regarding water: the functions of water, factors that influence water intake, the sources of water, how water is lost from the body, the recommended intake for water, factors that regulate blood volume and blood pressure, the electrolyte anions/cations, factors that regulate electrolytes, and factors that regulate acid-base balance.

Water	
Functions of Water	1. 2. 3. 4.
Factors that Influence Intake	1. 2. 3. 4.
Sources of Water	1. 2. 3. 4.

—Continued on next page

Fill in the Table for Water—*Continued*

Water	
How Water Is Lost	1. 2. 3. 4.
Recommendations for Intake	1. 2. 3. 4.
Factors that Regulate Blood Volume/Pressure	1. 2. 3. 4.
Electrolytes: Cations and Anions (Extra- and Intracellular)	1. 2. 3. 4.
Factors that Regulate Electrolytes	1. 2. 3. 4.
Factors that Regulate Acid-Base Balance	1. 2. 3. 4.

GETTING IT TOGETHER

Fill in the Table of Major Minerals

These tables may be used as study aids, submitted as part of course requirements, or used as extra credit opportunities as determined by your instructor. You may wish to enlarge the tables before completing them.

In the space provided, list two to four of the following: the major functions, deficiency disease and/or symptoms, toxicity symptoms, and the major food sources *found in your diet* for the major minerals. Refer to your Diet Analysis Project for your food sources. If a characteristic does not apply, write NA in the space.

For minerals not analyzed in your Diet Analysis Project, list appropriate food sources.

Name of Mineral	Function(s) of Mineral	Deficiency Disease and/or Symptoms	Toxicity Symptoms	Major Food Sources in Your Diet
Sodium	1. 2. 3. 4.	1. 2. 3. 4.	1. 2. 3. 4.	1. 2. 3. 4.
Chloride	1. 2. 3. 4.	1. 2. 3. 4.	1. 2. 3. 4.	1. 2. 3. 4.

—Continued on next page

Fill in the Table of Major Minerals—*Continued*

Name of Mineral	Function(s) of Mineral	Deficiency Disease and/or Symptoms	Toxicity Symptoms	Major Food Sources in Your Diet
Calcium	1. 2. 3. 4.	1. 2. 3. 4.	1. 2. 3. 4.	1. 2. 3. 4.
Phosphorus	1. 2. 3. 4.	1. 2. 3. 4.	1. 2. 3. 4.	1. 2. 3. 4.
Magnesium	1. 2. 3. 4.	1. 2. 3. 4.	1. 2. 3. 4.	1. 2. 3. 4.
Sulfur	1. 2. 3. 4.	1. 2. 3. 4.	1. 2. 3. 4.	1. 2. 3. 4.

PRACTICE TEST

The following items will help you check your understanding of this lesson. Compare your answers to the Answer Key at the end of the lesson. Review the course materials related to any incorrect answer.

Multiple Choice: Select the one choice that best answers the question.

1. Which of the following contributes most to the weight of the human body?
   A. Iron
   B. Water
   C. Protein
   D. Calcium

2. What organ provides the major control for homeostasis of body fluids?
   A. Liver
   B. Heart
   C. Kidneys
   D. Skeletal muscle

3. What is the major extracellular cation?
   A. Sodium
   B. Sulfate
   C. Protein
   D. Potassium

4. What is the term for the pressure that develops when two solutions of varying concentrations are separated by a membrane?
   A. Hypotension
   B. Hypertension
   C. Osmotic pressure
   D. Hypertonic pressure

5. What is another term for hypertension?
   A. High blood sodium
   B. High blood pressure
   C. Excessive mental stress
   D. Excessive muscular contraction

6. Almost all (99 percent) of the calcium in the body is used to __________
   A. provide energy for cells.
   B. provide rigidity for the bones and teeth.
   C. regulate the transmission of nerve impulses.
   D. maintain the blood level of calcium within very narrow limits.

7. Calcium absorption is facilitated by the presence of ________ in infants only.
   A. fiber.
   B. lactose.
   C. phytic acid.
   D. oxalic acid.

8. Sulfur is present in practically all __________
   A. vitamins.
   B. proteins.
   C. fatty acids.
   D. carbohydrates.

9. A person's bone density is highest at around age __________
   A. twenty-one years.
   B. thirty years.
   C. fifty-five years.
   D. seventy years.

10. Among the following ethnic groups, which has the highest bone density?
    A. Japanese
    B. Caucasian
    C. African American
    D. South American Hispanic

11. It would not be unusual for a person exercising for one hour to lose water weight in the amount of __________
    A. 10 to 15 pounds.
    B. 5 to 10 pounds.
    C. 4 to 8 pounds.
    D. 2 to 4 pounds.

12. Some of the physical symptoms reflecting dehydration in a child might include __________.

A. deep, dark circles under the eyes.
B. increased heart rate for age.
C. dry, tacky mucous membranes in the mouth.
D. all of the above.

13. Regarding sodium intake, an average adult only needs __________

A. 5 grams per day or one teaspoon.
B. 5,000 milligrams per day or one teaspoon.
C. 500 milligrams per day or one-tenth of a teaspoon.
D. A and B.

14. The most common cause(s) of osteoporosis is(are) __________

A. lack of estrogen in postmenopausal women.
B. increased bone formation in the elderly.
C. lack of weight-bearing exercise during the growing years.
D. all of the above.

15. What type of exercise can improve the outcome of osteoporosis?

A. Weight-bearing exercises, such as walking
B. Swimming laps or water aerobics
C. Meditation exercises
D. All of the above

16. An effect of increased water intake on the body of a person who is accustomed to greater quantities of salt is __________

A. reduced edema or swelling of ankles.
B. more bloating.
C. more energy.
D. all of the above.

Fill in the Blank: Insert the correct word or words in the blank for each item.

17. Fatigue is the first reliable sign that a person is ____________________.

18. A slow and steady loss of bone calcium resulting in fracturing or splintering of bones is a characteristic of ____________________.

Matching: Select the letter next to the word or phrase that best matches the numbered statements. Answers are used only once.

_____19. Most prevalent major mineral in the body

_____20. Crystalline structure of bone

_____21. Mineral that accounts for the structure of many proteins

A. Angiotensin
B. Calcium
C. Hydroxyapatite
D. Sodium
E. Sulfur

Essay: Reflect on the item(s) listed below; one or more may be included on the exam.

22. In what ways do the GI tract and the kidney function to help maintain fluid and electrolyte balance? How does the body defend itself when faced with conditions that induce excessive water and mineral losses (e.g., sweating, diarrhea)?

23. Explain the relationship between dietary sodium and hypertension. What are the roles of calcium, magnesium, and potassium in regulating blood pressure?

24. Discuss major risk factors in the development of osteoporosis. What population groups are most at risk? What dietary measures are advocated for high risk groups?

25. Discuss osteoporosis including a definition; how it develops; how age, sex, hormones, and genetics affect it; how activity affects it; and what dietary interventions can help prevent or treat osteoporosis.

ANSWER KEY

The following provides the answers and references for the Practice Test questions.

Answer		Learning Objective	Reference
1.	B	LO 1	p. 368
2.	C	LO 2	p. 371
3.	A	LO 1	p. 373
4.	C	LO 4	p. 374
5.	B	LO 3	p. 379
6.	B	LO 5	pp. 384–385
7.	B	LO 5	p. 386
8.	B	LO 5	p. 393
9.	B	LO 8	p. 389
10.	C	LO 8	p. 399
11.	D	LO 2	Video
12.	D	LO 7	Video
13.	C	LO 6	Video
14.	A	LO 8	Video
15.	A	LO 8	Video
16.	A	LO 4	Video
17.	dehydrated	LO 7	Video
18.	osteoporosis	LO 8	Video
19.	B	LO 5	p. 384
20.	C	LO 5	p. 385
21.	E	LO 5	p. 393
22.		LO 4	pp. 370–373
23.		LO 3, 5	pp. 379–380
24.		LO 8	pp. 396–400
25.		LO 8	Video

Lesson 15

Trace Minerals

LESSON ASSIGNMENTS

Text: Whitney and Rolfes, *Understanding Nutrition*, Chapter 13, "The Trace Minerals," pp. 403–433

Video: "Trace Minerals" from the series *Nutrition Pathways*

Project: "Diet Analysis Project"

Getting It Together:

The purpose of the "Getting It Together" section is to provide an outline of *new* or *difficult* material presented in the current lesson. This section provides practice in recalling lesson material and utilizes the textbook interactively. Use this section as a learning aid, to reinforce knowledge, or to organize information introduced in the lesson.

Related Activity:

This activity is not required unless your instructor assigns it. It is offered as a suggestion to help you learn more about the material presented in this lesson. Refer to your syllabus to determine whether this activity has been assigned.
Complete the attached form for the following related activity, and return it to your instructor according to the established deadline.
❐ "Calculating Iron Absorbed from Food"

Optional Web Activities:

Consult your instructor and/or syllabus for any assigned activities.

OVERVIEW

When you studied the major minerals, you learned that they are "major" because they are the minerals present in our bodies in the greatest amounts. Because we need them in such small quantities, the trace minerals are not enough to fill a teaspoon. Without them, however, we would not be able to exist. In this lesson, you will learn about the minor minerals which have an established RDA (iron, zinc, iodine, and selenium) and those which have ranges for safe and adequate daily intakes (copper, manganese, fluoride, chromium, and molybdenum). You will identify each minor mineral according to the roles it plays in healthy functioning; you will learn what happens when we are deficient in, or consume excessive amounts of, minor minerals; and you will learn what foods provide us with adequate amounts of minor minerals. In addition, you will delve more deeply into the functions of iron throughout pregnancy, adolescence, and in the aging adult as well as explore how iron-rich foods can be delivered to these populations.

LEARNING OBJECTIVES

Upon completing this lesson, you should be able to:

1. Explain the roles of iron in the body.
2. Describe how iron is absorbed, including the differences between heme and nonheme iron, and the factors affecting iron absorption.
3. Explain how iron is stored, transferred, and recycled in the body.
4. Describe the effects of iron deficiency and toxicity and the sources of iron for the body.
5. Describe the roles of zinc in the body.
6. Explain how zinc is absorbed, including factors affecting absorption.
7. Describe how zinc is transported and how it interacts with iron and copper.
8. Describe the effects of zinc deficiency and toxicity and the sources of zinc for the body.

9. Identify iodine, selenium, copper, manganese, fluoride, chromium, and molybdenum according to the following criteria:
   - Roles in the body
   - Deficiency symptoms and/or disease
   - Toxicity symptoms and/or disease
   - Sources for the body
10. Explain how the heavy metal lead impairs health.
11. Describe the benefits of phytochemicals for disease prevention.

TEXT FOCUS POINTS

The following focus points are designed to help you get the most from your reading. Review them, then read the assignment. You may want to write notes to reinforce what you have learned.

1. In what amount do trace minerals appear in the body? What are iron's two ionic states? Explain how iron works with protein in the electron transport chain. Where is most of the body's iron found? What is the function of iron in hemoglobin and myoglobin? Explain the function of mucosal ferritin, mucosal transferrin, and blood transferrin.
2. Define *heme* and *nonheme* iron. What foods contain heme and nonheme iron? What percent of iron consumed by an average person in a day comes from heme sources? At what rate is heme iron absorbed? At what rate is nonheme iron absorbed?
3. What are the factors that enhance nonheme iron absorption? What are the factors that inhibit iron absorption? What three factors are most relevant when analyzing iron content of meals? Explain how iron absorption is impacted by health status, stage in the life cycle, and iron status. Where is surplus iron stored in the body and how is it stored? What is hemosiderin? Explain how iron is recycled in the body.
4. How prevalent is iron deficiency? In what ways can blood, and therefore, iron, be lost from the body? What populations are at high risk for iron deficiency? Describe the three stages of iron deficiency. Define *iron deficiency* and *iron*

deficiency anemia. What are symptoms of iron deficiency anemia? How does iron deficiency affect behavior? What is pica?

5. What is hemochromatosis? How prevalent is hemochromatosis? What is hemosiderosis? List the diseases that can be aggravated by hemochromatosis. What are symptoms of iron overload? What population is most prone to iron overload? How does iron intake relate to the following: heart disease, cancer, and poisoning?

6. What foods contribute the most iron? How do contamination iron and supplemental iron contribute to intakes?

7. Where are the highest concentrations of zinc in the body? What are the roles of zinc in the body?

8. What factors limit zinc bioavailability? Define the terms *metallothionein* and *enteropancreatic circulation.*

9. What is the main transport vehicle for zinc? How does zinc absorption affect iron and copper status?

10. What are the symptoms of zinc deficiency? What form of retardation is associated with zinc deficiency? What are the effects of zinc toxicity with regard to zinc supplements? What foods are the richest sources of zinc?

11. What are the major roles of iodide in the body? What condition develops with an iodine deficiency? What is a goitrogen? What is cretinism? What condition develops with iodine toxicity? What is the major source of the world's iodine? What process eliminated iodine deficiency during the 1930s in the United States and Canada? What foods contribute to iodine intake?

12. What is the role of selenium in the body? Define *glutathione peroxidase.* What condition found in children in China is associated with selenium deficiency? What food sources provide the most selenium? List the symptoms of selenium toxicity.

13. What function does copper serve in the body? What two minerals can interfere with copper absorption? What are the best food sources of copper?

14. How does manganese function in the body? What other minerals may inhibit manganese absorption?

15. What is the function of fluoride in the body? What is fluorapatite? What is the best source of fluoride in the diet? What condition results from fluoride deficiency? Define *fluorosis*.

16. How does chromium function in the body? What are the best sources of chromium? What is the controversy surrounding chromium supplements? What job does molybdenum perform in the body? What are the best food sources of molybdenum?

17. Explain the adverse effects of lead poisoning as an environmental contaminant.

VIDEO FOCUS POINTS

The following focus points are designed to help you get the most from the video segment of this lesson. Review them before watching the video. You may want to write notes to reinforce what you have learned.

1. What is the most important nutrient deficiency? Which populations are at risk for iron deficiency? What causes iron deficiency? What are symptoms of iron deficiency? Which foods are high in iron? What enhances the absorption of non-heme iron?

2. What is hereditary hemochromatosis and how is it detected? What is the impact on health if hemochromatosis is not detected? Which populations are at greater risk? What is the treatment for hemochromatosis?

3. What is the role of chromium? What are good food sources of chromium? What controversy surrounds chromium supplements and diabetes management?

4. How does fluoride function in the body? What are good sources of fluoride? What is the link between fluoride and dental health?

5. What are phytochemicals? What foods are good sources of phytochemicals? Explain the link between lycopene and prostate cancer. What are good sources of lycopene?

6. How could iron intake impact someone with a history of heart disease?

PROJECT

This project is required in order to complete the course successfully.

This portion of the lesson is designed to provide you with specific information about nutrient adequacy and balance, number of kcalories consumed, nutrient density, and variety of the foods you typically eat. After analyzing a three-to-seven-day food intake, you will be able to apply the information you have learned to make appropriate changes to improve eating habits.

Refer to the Diet Analysis Project for complete details and directions. Contact your instructor if you have any questions. The Diet Analysis Project is in the back of this student course guide.

Information provided in Chapter 13 will be useful when answering questions pertaining to your Diet Analysis Project with regard to minor minerals.

GETTING IT TOGETHER

Fill in the Table of Trace Minerals

These tables may be used as study aids, submitted as part of course requirements, or used as extra credit opportunities as determined by your instructor. You may wish to enlarge the tables before completing them.

In the space provided, list two to four of the following: the major functions, deficiency disease and/or symptoms, toxicity symptoms, and the major food sources *found in your diet* for the trace minerals. Refer to your Diet Analysis Project for your food sources. If a characteristic does not apply, write NA in the space.

For minerals not analyzed in your Diet Analysis Project, list appropriate food sources.

Name of Mineral	Function(s) of Mineral	Deficiency Disease and/or Symptoms	Toxicity Symptoms	Major Food Sources in Your Diet
Iron	1. 2. 3. 4.	1. 2. 3. 4.	1. 2. 3. 4.	1. 2. 3. 4.
Zinc	1. 2. 3. 4.	1. 2. 3. 4.	1. 2. 3. 4.	1. 2. 3. 4.
Iodine	1. 2. 3. 4.	1. 2. 3. 4.	1. 2. 3. 4.	1. 2. 3. 4.

—Continued on next page

Fill in the Table of Trace Minerals—*Continued*

Name of Mineral	Function(s) of Mineral	Deficiency Disease and/or Symptoms	Toxicity Symptoms	Major Food Sources in Your Diet
Selenium	1. 2. 3. 4.	1. 2. 3. 4.	1. 2. 3. 4.	1. 2. 3. 4.
Copper	1. 2. 3. 4.	1. 2. 3. 4.	1. 2. 3. 4.	1. 2. 3. 4.
Manganese	1. 2. 3. 4.	1. 2. 3. 4.	1. 2. 3. 4.	1. 2. 3. 4.
Fluoride	1. 2. 3. 4.	1. 2. 3. 4.	1. 2. 3. 4.	1. 2. 3. 4.
Chromium	1. 2. 3. 4.	1. 2. 3. 4.	1. 2. 3. 4.	1. 2. 3. 4.

Calculating Iron Absorbed from Food

NAME: ______________________________ ID #: ______________________________

This activity will help you determine how much iron is found in foods and how much is absorbed into the body.

Refer to Chapter 13 for information on heme iron and nonheme iron to answer the questions below.

DIRECTIONS:

1. Use diet analysis software or Appendix H in your text to calculate the amount of iron and vitamin C found in the following foods.
2. Respond to the questions that follow.

FOOD	Mg Iron	Mg Vitamin C
3 oz chicken breast meat		
⅔ cup cooked broccoli		
1 small baked potato		
2 tbs sour cream		
1 cup vanilla ice cream		
TOTALS		

1. ________ How many milligrams (mg) total iron was in the chicken?

2. ________ On the average, 40 percent heme iron is found in chicken. Calculate the following to determine mg of *heme iron* in the chicken.

 _____ mg (Step 1) x 0.40 = _____ mg heme iron

3. ________ How much total iron came from the broccoli, potato, sour cream, and yogurt? (*Show your calculations.*)

4. ________ Calculate the following to obtain mg of *nonheme* iron.

 _____mg (Step 1) x 0.60 + _____ mg (Step 3) = _____ mg

—Continued on the back

Calculating Iron Absorbed from Food—*Continued*

5. ________ How much vitamin C was in the above meal? Use the following information:
   - Less than 25 mg = low
   - 25–75 mg = medium
   - more than 75 mg = high

6. ________ What is the percent rate of heme iron absorbed?

7. ________ Calculate the following to obtain total mg of heme iron absorbed.

   _____ mg (Step 2) x _____% (Step 6) = _____ mg

8. ________ Use the following information to determine how much MFP factor was in the chicken:
   - Less than 1 ounce chicken = low
   - 1 to 3 ounces chicken = medium
   - more than 3 ounces = high MFP

9. ________ Review the amounts of vitamin C (Step 5) and MFP (Step 8) to determine the percent availability of nonheme iron. Use the following information:
   - If either vitamin C or MFP was high or if both were medium, the availability is high or 8 percent
   - if neither was high, but one was medium, the availability is medium or 5 percent
   - if both were low, availability is low or 3 percent.

10. ________ Complete the following to determine the actual amount of mg of nonheme iron absorbed.

   _____ mg (Step 4) x _____% availability (Step 9) =_____ mg

—*Continued on next page*

Calculating Iron Absorbed from Food—*Continued*

11. ___________ Add mg heme absorbed (Step 7) and mg nonheme absorbed (Step 10) to obtain total iron absorbed from the meal. (*Show calculations.*)

12. ___________ What is your RDA for iron (*see the RDA Table in the text or use diet analysis software.*)?

13. ___________ The RDA for iron is based on the assumption that you will absorb 10 percent of the iron you ingest. Therefore, if you are a man of any age or a woman over fifty, you need to absorb one mg per day of iron. If you are a woman age eleven to fifty, you need to absorb 1.5 mg per day of iron. To determine what percent of your iron needs would be met with the above meal, divide 1 mg or 1.5 mg into mg iron absorbed (Step 12). *(Show calculations.)*

14. Would the percent of iron obtained from the meal above be enough to satisfy your DAILY personal RDA for iron? If not, how much more iron would you need to meet your personal RDA?

15. What, if anything, could you do to improve your iron status? (*Cite page numbers in the text to support your answer.*)

PRACTICE TEST

The following items will help you check your understanding of this lesson. Compare your answers to the Answer Key at the end of the lesson. Review the course materials related to any incorrect answer.

Multiple Choice: Select the one choice that best answers the question.

1. What iron-containing compound carries oxygen in the bloodstream?
   A. Ferritin
   B. Myoglobin
   C. Transferrin
   D. Hemoglobin

2. Absorption of iron from supplements is improved by taking them ________
   A. with tea.
   B. with orange juice.
   C. with milk.
   D. between meals on an empty stomach.

3. Which of the following characteristics is shared by zinc and iron?
   A. Good food sources include dairy products.
   B. Proteins in the blood are needed for their transport.
   C. Severe deficiencies lead to delay in the onset of puberty.
   D. Doses of ten times the RDA are known to cause death in children.

4. Which of the following population groups is LEAST susceptible to iron deficiency anemia?
   A. Older infants
   B. Children two to ten years of age
   C. Women of childbearing age
   D. Men twenty to forty-five years of age

5. What is the name given to the ingestion of nonnutritive substances?
   A. Pica
   B. Goiter
   C. Tetany
   D. Hemosiderosis

6. Which of the following is known to enhance iron absorption?
   A. Tea
   B. Coffee
   C. Foods containing vitamin C
   D. Foods containing vitamin E

7. Zinc is known to play an important role in all of the following functions EXCEPT __________
   A. wound healing.
   B. synthesis of retinal.
   C. production of sperm.
   D. oxidation of polyunsaturated fatty acids.

8. Zinc is highest in foods that also contain a high amount of __________
   A. fat.
   B. fiber.
   C. protein.
   D. carbohydrate.

9. Which of the following is a property of selenium in nutrition?
   A. It participates in the functioning of insulin.
   B. Severe deficiency is associated with heart disease in China.
   C. Significant food sources include dairy and unprocessed vegetables.
   D. It has no RDA, but the estimated safe and adequate dietary intake is only 2–3 µg/day.

10. Chromium deficiency is characterized by __________
    A. hypertension.
    B. hyperglycemia.
    C. enlargement of the liver.
    D. enlargement of the thyroid gland.

11. Which of the following is known to enhance iron absorption?
    A. Tea
    B. Coffee
    C. Vitamin D
    D. Vitamin C

12. Which of the following is a characteristic of iron deficiency?
    A. Blood erythrocyte protoporphyrin levels decline as anemia worsens.
    B. Iron supplements are not as effective in treating anemia as proper nutrition is.
    C. People with anemia generally become fatigued only when they exert themselves.
    D. The concave nails of iron deficiency anemia result from abnormal ferritin levels.

13. Iron overload is also known as __________
    A. ferrocyanosis.
    B. hemoglobinemia.
    C. hemochromatosis.
    D. metalloferrothionosis.

14. The most common cause of iron overload is __________
    A. an injury to the GI tract.
    B. a genetic predisposition.
    C. excessive use of iron cookware.
    D. excessive use of iron supplements.

15. Which of the following mechanisms explains why fluoride is effective in controlling tooth decay?
    A. It helps regulate calcium levels in saliva.
    B. It helps form decay-resistant flurapatite.
    C. It inhibits growth of decay-producing bacteria.
    D. It changes the pH of the mouth, inhibiting bacterial growth.

16. What is the most reliable source of dietary fluoride?
    A. Public water
    B. Dark green vegetables
    C. Milk and milk products
    D. Meats and whole grain cereals

17. Which of the following is a characteristic of chromium in nutrition?
    A. A chromium deficiency leads to hypothyroidism.
    B. Chromium supplements are proven to be helpful.
    C. Whole grains represent an excellent source.
    D. In the body, chromium enhances the action of ceruloplasmin.

Matching: Select the letter next to the word or phrase that best matches the numbered statements. Answers are used only once.

_____18. Form of iron found only in animal flesh

_____19. Zinc-binding protein of intestine

_____20. Iodine deficiency disease

A. Tea
B. Heme
C. Goiter
D. Metallothionein
E. Anemia

Essay: Reflect on the item(s) listed below; one or more may be included on the exam.

21. Explain the difference between heme and nonheme iron. How can the efficiency of absorption be increased for both types of iron?

22. In the proper sequence, describe the three stages in the development of iron deficiency.

23. Discuss the effects of lead exposure on health and human performance.

ANSWER KEY

The following provides the answers and references for the Practice Test questions.

	Answer	Learning Objective	Reference
1.	D	LO 1	textbook, p. 406
2.	D	LO 2	textbook, p. 414
3.	B	LO 5, 7	textbook, p. 415
4.	D	LO 4	textbook, p. 409
5.	A	LO 4	textbook, p. 410
6.	C	LO 2	textbook, p. 413
7.	D	LO 5	textbook, pp. 414–415
8.	C	LO 6, 8	textbook, p. 416
9.	B	LO 9	textbook, p. 419
10.	B	LO 9	textbook, p. 423
11.	D	LO 2	Video
12.	C	LO 4	Video
13.	C	LO 4	Video
14.	B	LO 4	Video
15.	B	LO 9	Video
16.	A	LO 9	Video
17.	C	LO 9	Video
18.	B	LO 2	textbook, p. 406
19.	D	LO 6	textbook, p. 415
20.	C	LO 9	textbook, p. 418
21.		LO 2	textbook, pp. 406–408
22.		LO 4	textbook, pp. 409–410
23.		LO 10	textbook, p. 424

Lesson 16

Physical Activity: Fitness Basics

LESSON ASSIGNMENTS

Text: Whitney and Rolfes, *Understanding Nutrition*, Chapter 14, "Fitness: Physical Activity, Nutrients, and Body Adaptations," pp. 437–444

Video: "Physical Activity: Fitness Basics" from the series *Nutrition Pathways*

Related Activity:

This activity is not required unless your instructor assigns it. It is offered as a suggestion to help you learn more about the material presented in this lesson. Refer to your syllabus to determine whether this activity has been assigned.

Complete the attached form for the following related activity, and return it to your instructor according to the established deadline.

❒ "Designing a Personal Fitness or Health Program"

Optional Web Activities:

Consult your instructor and/or syllabus for any assigned activities.

OVERVIEW

Although the primary focus of this course is the acquisition of nutrition knowledge leading toward a healthier life, the equation is only partially completed: the inclusion of physical activity as a regular part of one's life provides one more component of the equation leading toward total well-being. This lesson demonstrates clearly how physical activity can have a significant, positive influence on your overall health, whether you are a seasoned veteran of activity, a novice, or someone who has never been very active. Combined with appropriate nutrition choices and other positive lifestyle changes, physical activity is nearly the "magic bullet" we all seem to be seeking for ultimate health and well-being. You will learn how to incorporate your personal goals into a program that emphasizes more activity in your life, whether or not those goals are health- or fitness-related. You will also debunk the myths associated with physical activity by learning what fuels are best suited for specific activities, and finally, you will learn to design a program appropriate for you regardless of your age, sex, financial status, or health.

LEARNING OBJECTIVES

Upon completing this lesson, you should be able to:

1. State the guidelines for developing and maintaining fitness.

2. State the guidelines for obtaining health.

3. Define *fitness*.

4. Cite general research findings and statistics surrounding participation in regular physical activity.

5. Cite the benefits associated with being more active.

6. List the components of a sound fitness program.

7. Explain how activity and sound nutrition can be incorporated into one's life, whether at work or at home.

TEXTBOOK FOCUS POINTS

The following focus points are designed to help you get the most from your reading. Review them, then read the assignment. You may want to write notes to reinforce what you have learned.

1. What is the main purpose of the American College of Sports Medicine (ACSM) guidelines for physical fitness? What are the ACSM guidelines for developing and maintaining physical fitness?

2. What are the guidelines for obtaining health benefits?

3. Provide three definitions of fitness. Define *sedentary*.

4. What does research state about regular physical activity? What percent of the U.S. adult population exercises regularly? What degenerative diseases are directly linked to the lack of physical activity? How does being physically inactive relate to early death? State the ACSM advice regarding public health effort focus.

5. Explain how being more active impacts the following: sleep, nutritional health, body composition, bone density, immunity, risk for cancer, circulation and lung function, risk for cardiovascular disease, risk for type 2 diabetes, gallbladder disease, anxiety and depression, self-image, and longevity.

6. Define the following components of a sound fitness program: flexibility, muscle strength, muscle endurance, and cardiorespiratory endurance. Explain the following principles of conditioning: progressive overload, frequency, intensity, duration, warm-up, and cool-down. Describe how the ACSM classifies individuals based on coronary risk factors. Define the following terms: *hypertrophy*, *atrophy*, *moderate exercise*, and *resistance training*.

7. What are the benefits of resistance training? What are the benefits of aerobic training for cardiorespiratory endurance? For muscle conditioning? Provide an example of a balanced fitness program.

VIDEO FOCUS POINTS

The following focus points are designed to help you get the most from the video segment of this lesson. Review them before watching the video. You may want to write notes to reinforce what you have learned.

1. What populations are at increased risk for dying from all causes?

2. What does the latest research state regarding ultra-athletes or marathoners and immune function?

3. What advantages does an on-site fitness program provide to employees? How does exercise affect brain chemistry?

4. Describe a general but adequate diet to support activity.

5. Why is walking considered one of the best types of exercise to perform? What exercise factors positively impact health and longevity the most?

6. What are three preliminary steps to take prior to beginning an exercise program? Why would a person with high blood cholesterol consider participating in a regular, aerobic exercise program?

Designing a Personal Fitness or Health Program

NAME: ______________________________ ID #: ________________

The following activity will assist you in developing an activity program and eating plan that is best suited to your personal goals and lifestyle.

Refer to Chapter 14 in the textbook to complete the following activity.

1. In which of the following are you most interested (check one)?

   ____ Becoming physically fit
   ____ Becoming healthy
   ____ Becoming both fit and healthy
   ____ Maintaining current condition

2. What best describes your current activity program?

   I exercise:
   ____ Daily
   ____ At least three times a week
   ____ One to two times a week
   ____ Once a month
   ____ Every two to three months
   ____ Rarely

   Type of Activity:
   ____ Walk/jog/run
   ____ Hike/cycle
   ____ Swim
   ____ Play tennis/golf
   ____ Garden
   ____ Other (please list below)

3. How much time do you spend participating in selected activities?

   I exercise for:
   ____ Twenty to sixty minutes, continuously
   ____ Up to thirty minutes, intermittently (that is, throughout the day)

—Continued on the back

Cite page numbers in your text to support your responses to the following questions:

4. What are the ACSM guidelines for achieving fitness versus health?

   Fitness:

   Health:

5. Describe a *detailed* nutrition program that would support fitness or health goals.

PRACTICE TEST

The following items will help you check your understanding of this lesson. Compare your answers to the Answer Key at the end of the lesson. Review the course materials related to any incorrect answer.

Multiple Choice: Select the one choice that best answers the question.

1. According to the American College of Sports Medicine, which of the following would meet the exercise schedule to improve cardiorespiratory endurance?
   A. Two hours of aerobic exercise daily
   B. One hour of strength training four times a week
   C. Five minutes of aerobic exercise two times a week
   D. Thirty minutes per day of any continuous activity using large muscle groups

2. A muscle that increases size in response to use is an example of ______
   A. atrophy.
   B. hypertrophy.
   C. muscular endurance.
   D. muscle engorgement.

3. The effect of regular exercise on heart and lung function is known as ________
   A. muscle fitness.
   B. muscle endurance.
   C. cardiopulmonary adaptation.
   D. cardiorespiratory conditioning.

4. Research clearly shows that the risk of dying not only from heart disease but from all causes is greatest for ________
   A. elite athletes such as marathoners.
   B. sedentary people, also known as "couch potatoes."
   C. occasional exercisers, also known as "weekend warriors."
   D. moderately active people who exercise three to four days a week.

5. Individuals who run sixty or seventy miles per week, such as marathon runners, may actually __________
   A. feel it is not enough to achieve their goals.
   B. be overexercising to the benefit of their health.
   C. suppress their immune function and cause disease.
   D. all of the above.

6. In addition to being able to exercise at work, an employee who has access to a wellness program and fitness facility at the job site __________
   A. is more likely to make positive changes in lifestyle.
   B. has less absenteeism from work.
   C. becomes more educated in health issues.
   D. all of the above.

7. Regardless of whether the intensity of an exercise program is low, moderate, or vigorous, the best diet emphasizes __________
   A. protein.
   B. fat.
   C. carbohydrate.
   D. water.

8. Walking is considered the best form of exercise because it __________
   A. is not expensive.
   B. promotes cardiovascular endurance.
   C. is easy to perform.
   D. all of the above.

9. A person with high blood cholesterol should consider participating in an aerobic exercise program such as walking or jogging because aerobic exercise __________
   A. will raise HDL ("good") cholesterol after a period of training and improve the risk for heart problems.
   B. will cause a sudden increase in the heart rate to near maximal levels which lowers the "bad" cholesterol.
   C. forces more LDL ("bad") cholesterol out of the body through waste elimination.
   D. keeps the blood circulating faster, even at rest, therefore causing a decrease in total blood cholesterol.

Matching: Select the letter next to the word or phrase that best matches the numbered statements. Answers are used only once.

_____10. The capacity of the joints to move with less chance of injury

_____11. Increase in muscle size and strength

_____12. Required for aerobic metabolism

A. Hypertrophy
B. Oxygen
C. Flexibility
D. Muscle engorgement

Essay: Reflect on the item(s) listed below; one or more may be included on the exam.

13. Discuss the physiological and psychological benefits of being physically fit.

14. Explain the recommended training procedure (i.e., overload principle) for mastering the components of fitness.

ANSWER KEY

The following provides the answers and references for the Practice Test questions.

	Answer	Learning Objective	Reference
1.	D	LO 1	textbook, p. 440
2.	B	LO 3	textbook, p. 441
3.	D	LO 5	textbook, p. 440
4.	B	LO 2	Video
5.	C	LO 5	Video
6.	D	LO 7	Video
7.	C	LO 6	Video
8.	D	LO 3, 7	Video
9.	A	LO 2, 5	Video
10.	C	LO 6	textbook, p. 440
11.	A	LO 5	textbook, p. 441
12.	B	LO 4	textbook, p. 439
13.		LO 5	textbook, pp. 438–439
14.		LO 6	textbook, pp. 440–444

Lesson 17

Physical Activity: Beyond Fitness

LESSON ASSIGNMENTS

Text: Whitney and Rolfes, *Understanding Nutrition*, Chapter 14, "Fitness: Physical Activity, Nutrients, and Body Adaptations," pp. 444–466

Video: "Physical Activity: Beyond Fitness" from the series *Nutrition Pathways*

Optional Web Activities:
Consult your instructor and/or syllabus for any assigned activities.

OVERVIEW

You have examined the differences between "fitness" and "health" as they relate to activity; you differentiated between the components of a sound fitness program; and you looked at the benefits associated with being more physically active—all of these topics emphasized the very basics of physical activity. Lesson 17 goes beyond the basic information you need to begin being more active. In this lesson, you will learn how the body uses food to fuel activities, how specific nutrients support activity, how water and electrolytes affect the athlete, what is an acceptable diet for a physically active person, and how sports drinks and other ergogenic aids benefit or hinder athletic performance.

LEARNING OBJECTIVES

Upon completing this lesson, you should be able to:

1. Describe how ATP and CP are used in physical activity.
2. Explain how glucose is used during physical activity.
3. Explain how fat is used during physical activity.

4. Explain how protein is used during physical activity and between times of activity.

5. Describe the use of vitamins, minerals, and supplements during activity.

6. State how fluids and electrolytes impact activity.

7. Describe a diet to support fitness as well as a diet suitable for pre- and post-competition.

8. Explain the effects of ergogenic aids on athletic performance.

9. Explain the importance of diet and exercise for physically challenged people and people with high blood lipids.

TEXT FOCUS POINTS

The following focus points are designed to help you get the most from your reading. Review them, then read the assignment. You may want to write notes to reinforce what you have learned.

1. What compounds are the driving force for muscle contraction? What is creatine phosphate? During rest, how much of the body's energy comes from fatty acids? What factors affect the type of fuel used by muscles during activity? Define *anaerobic* and *aerobic*.

2. What happens to muscles when glycogen is depleted? How do high-carbohydrate diets impact endurance activities such as running? What is the primary fuel source for activities of extremely intense, short duration? How soon will glycogen depletion occur after the onset of intense activity? During moderate physical activity, what fuels are used? How does intense activity impact fuel use? What is lactic acid? What is the Cori cycle?

3. What is the primary fuel used during the first twenty minutes of moderate activity? What is the primary fuel used after twenty minutes of moderate activity? What do marathon runners refer to as the point of total glucose depletion? What steps can endurance athletes take to maximize glucose supply? What does eating high-carbohydrate foods after physical activity do to glycogen stores? How does training affect glycogen stores? What is "carbohydrate loading"? What is the effect of a high-carbohydrate meal eaten

within fifteen minutes after activity? What is the primary fuel source for untrained muscles?

4. From what areas of the body are fatty acids used during activity? What is the source of fat during the early phase of activity? What hormone is released during exercise to help break down fat? What happens to fat utilization as exercise intensity increases? What are the factors that influence fat use during activity? How does training affect fat use? What advice is given to people who want to control weight and lose fat with regard to exercise intensity and duration? What is the bottom line regarding activity and weight and/or fat loss? Explain the "rule of thumb" regarding breathing and exercise intensity.

5. What happens to body protein synthesis during activity? During active muscle-building training, how much muscle may an athlete add each day to body protein? How much protein is typically used as a fuel source? What are the factors that affect protein use during activity? Explain how the following factors affect protein use during exercise: diet, intensity, duration, and training. How much protein is recommended for active people and athletes? What dietary components should athletes increase?

6. Explain whether athletes need nutrient supplements and, if so, which ones. What nutrient deficiency is a potential risk for female athletes? What is "sports anemia"?

7. What is the first symptom of dehydration in an athlete? What percent water loss can reduce the capacity for muscles to work? What percent water loss will cause a person to collapse? How does the body cool itself? Differentiate between hyperthermia, heat stroke, and hypothermia. State the hydration schedule for physical activity. What is the best fluid for an everyday, active person? For endurance athletes? Why drink "cool" water? What is hyponatremia? How can athletes prevent hyponatremia? What are the advantages athletes receive from consuming sports drinks? Why should athletes be concerned about caffeine and alcohol intake?

8. How much water should you drink to replace one pound of weight lost during exercise? What is the basis of a healthy diet for an active person? How many kcalories are recommended for a pregame meal? How many hours before competition should the meal be eaten? What foods are recommended for a pregame meal? What foods are recommended after competition?

VIDEO FOCUS POINTS

The following focus points are designed to help you get the most from the video segment of this lesson. Review them, then watch the video. You may want to write notes to reinforce what you have learned.

1. What substance is depleted during repetitive exercises? What foods should be the foundation of a diet for an athlete? How much food should be eaten the day of a competition? Why are high-carbohydrate foods recommended the day before a competition? What advice is given regarding caffeine intake prior to exercise?

2. Explain how exercise and nutrition impact a physically challenged person.

3. What are ergogenic aids? How do nutrition supplements impact the performance of athletes? What are some of the side effects of steroid use among athletes? What is epoetin? What advice is given regarding the use of sports drinks and performance?

4. How can regular exercise impact energy, stress, and blood lipid levels?

PRACTICE TEST

The following items will help you check your understanding of this lesson. Compare your answers to the Answer Key at the end of the lesson. Review the course materials related to any incorrect answer.

Multiple Choice: Select the one choice that best answers the question.

1. Which of the following is a property of conditioned muscles?
   A. They can store more glycogen.
   B. They are more efficient at converting fat to glucose.
   C. They contain less mitochondria due to increased glucose utilization.
   D. They rely less on fat breakdown and more on glucose oxidation for energy.

2. Which of the following describes fat utilization during physical activity?
   A. Fat that is stored closest to the exercising muscle is oxidized first.
   B. Fat oxidization makes more of a contribution as the intensity of the exercise increases.
   C. Fat oxidation may continue at an above-normal rate for some time after cessation of physical activity.
   D. Fat is burned in higher quantities during short high-intensity exercises than prolonged low-intensity exercises.

3. Which of the following is a role for diet in physical activity?
   A. Diets high in fat lead to a fall in amino acid utilization for fuel.
   B. Diets lacking in carbohydrates lead to increased amino acid utilization for fuel.
   C. Deficiencies of vitamins have no effect on performance, provided that all other nutrients are adequate.
   D. Deficiencies of minerals have no effect on performance, provided that all other nutrients are adequate.

4. Which of the following describes the role of protein in the diet of competitive athletes?
   A. The need for protein per kilogram body weight is higher in females than in male athletes.
   B. The need for protein is best met by increasing the level to 20 to 25 percent of total energy content of the diet.
   C. The need for protein in weight lifters and marathon runners may be up to 75 to 100 percent higher than the RDA.
   D. The need for protein in most athletes generally would NOT be sufficient from diets meeting energy requirements but containing only 10 percent of the energy as protein.

5. Which of the following is a known feature of iron nutrition in athletes?
   A. Iron in sweat represents the major route of iron loss from the body.
   B. Iron deficiency affects a higher percentage of male athletes than female athletes.
   C. Sports anemia is successfully treated by increasing dietary iron to levels two to three times the RDA.
   D. Iron losses in runners occur when blood cells are squashed by the impact of the foot on a hard surface.

6. The first sign of dehydration is typically __________
   A. fatigue.
   B. dizziness.
   C. intense thirst.
   D. intense sweating.

7. Which of the following is a feature of water metabolism during exercise?
   A. The maximum loss of fluid per hour of exercise is about 0.5 liters.
   B. In cold weather, the need for water falls dramatically because the body does not sweat.
   C. Sweat losses can exceed the capacity of the GI tract to absorb water, resulting in some degree of dehydration.
   D. Heavy sweating leads to a marked rise in the thirst sensation to stimulate water intake which delays the onset of dehydration.

8. Which of the following would be the best choice for physically active people who need to rehydrate?
   A. "Sweat" replacers
   B. Salt tablets and tap water
   C. Diluted juice or cool water
   D. Water warmed to body temperature

9. What is the recommended composition of the postgame meal of the athlete?
   A. Low protein
   B. High protein
   C. Low carbohydrate
   D. High carbohydrate

10. Whether performing an anaerobic sport, such as diving, or an aerobic sport, such as marathon running, the training routine __________
    A. involves repetitive use of muscles during the exercise.
    B. depletes muscle glycogen.
    C. must be supported by a high-carbohydrate diet.
    D. all of the above.

11. Regardless of the physical pursuits of an athlete, the recommended diet should be a foundation of __________
    A. protein.
    B. amino acids.
    C. water.
    D. complex carbohydrates.

12. Physical activity is very important to physically challenged people because ______
    A. it keeps them from becoming bored.
    B. they can eat more and, by consuming more nutrients, become healthier.
    C. they are not as strong as people who are not physically challenged.
    D. all of the above.

13. The illegal drug, used by athletes as an alternative for blood doping, is _________
    A. epoetin.
    B. an anabolic steroid.
    C. effective, as long as it is taken in moderation.
    D. all of the above.

Fill in the Blank: Insert the correct word or words in the blank for each item.

14. The foundation of an athlete's diet should consist of __________________.

15. The ____________ component of a blood lipid profile has been shown to be improved by regular aerobic exercises.

Matching: Select the letter next to the word or phrase that best matches the numbered statements. Answers are used only once.

_____16. Number of minutes after starting a physical activity for blood fatty acid concentrations to rise

_____17. The replacement of fluids during physical activities

_____18. Recommended carbohydrate intake, in grams per kilogram of body weight of athletes in heavy training

A. Eight
B. Fifteen
C. Twenty
D. Hydration
E. Metabolism

Essay: Reflect on the item(s) listed below; one or more may be included on the exam.

19. Discuss the use of protein, fat, and carbohydrate as fuels during low, moderate, and intense exercise.

20. Discuss the need for water in maintaining physical performance. What are the symptoms of dehydration? What are the recommendations for ensuring that the body is well-hydrated prior to an athletic event?

ANSWER KEY

The following provides the answers and references for the Practice Test questions.

	Answer	Learning Objective	Reference
1.	A	LO 1, 3	textbook, p. 447
2.	C	LO 3	textbook, pp. 448–449
3.	B	LO 2	textbook, pp. 450–451
4.	C	LO 4	textbook, pp. 450–451
5.	D	LO 5	textbook, pp. 452–453
6.	A	LO 6	textbook, p. 453
7.	C	LO 6	textbook, p. 453
8.	C	LO 6	textbook, p. 455
9.	D	LO 7	textbook, p. 459
10.	D	LO 1, 2	Video
11.	D	LO 7	Video
12.	B	LO 9	Video
13.	A	LO 8	Video
14.	complex carbohydrates	LO 7	Video
15.	HDL (or triglycerides)	LO 9	Video
16.	C	LO 1, 3	textbook, p. 449
17.	D	LO 6	textbook, pp. 453–454
18.	A	LO 7	textbook, p. 447
19.		LO 1, 2, 3, 4	textbook, pp. 446–450
20.		LO 6	textbook, pp. 453–457

Lesson 18

Life Cycle: Pregnancy

LESSON ASSIGNMENTS

Text: Whitney and Rolfes, *Understanding Nutrition*, Chapter 15, "Life Cycle Nutrition: Pregnancy and Lactation," pp. 469–491 and pp. 501–503

Video: "Life Cycle: Pregnancy" from the series *Nutrition Pathways*

Related Activity:

This activity is not required unless your instructor assigns it. It is offered as a suggestion to help you learn more about the material presented in this lesson. Refer to your syllabus to determine whether this activity has been assigned.

Complete the attached form for the following related activity, and return it to your instructor according to the established deadline.

❒ "Evaluating the Life Cycle"

Optional Web Activities:

Consult your instructor and/or syllabus for any assigned activities.

OVERVIEW

As you enter different stages of growth and development throughout the life cycle, your nutritional needs will vary. You will require the same nutrients as someone much older or much younger than yourself, but the amount needed will be unique to the stage of life. This lesson focuses on nutrition needs as a woman prepares for pregnancy, as well as how nutrition affects fetal growth and development. You will see how exercise plays a part as maternal needs change during pregnancy, how weight gain impacts pregnancy, and how teen pregnancy impacts the mother as well as the developing fetus.

LEARNING OBJECTIVES

Upon completing this lesson, you should be able to:

1. Briefly describe the function of the placenta.
2. Describe the developmental stage that defines an embryo and a fetus.
3. Cite the critical periods of fetal development.
4. Explain how weight affects pregnancy and how much weight should be gained.
5. Describe appropriate exercise guidelines during pregnancy.
6. State energy and nutrient needs during pregnancy.
7. Describe nutritional strategies for combating common pregnancy-related concerns.
8. Explain how malnutrition impacts pregnancy.
9. Explain how infant birthweight impacts the health of the infant.
10. Explain the impact of diabetes and hypertension on pregnancy.
11. Present risks associated with adolescent pregnancy and pregnancy for older women.
12. Specify how alcohol, drugs, and smoking impact pregnancy.
13. Describe circumstances that could interfere with personal fitness and nutritional goals after pregnancy.

TEXT FOCUS POINTS

The following focus points are designed to help you get the most from your reading. Review them, then read the assignment. You may want to write notes to reinforce what you have learned.

1. What is the function of the placenta? What attaches the placenta to the fetus?
2. What developments occur in the embryo at eight weeks after conception? What is a fetus?

3. What are “critical periods” during pregnancy? At what point during pregnancy do neural tube defects occur? What is anencephaly? What is spina bifida? List the risk factors for neural tube defects. What nutrient can prevent neural tube defects? When should a woman use folate supplementation? By what percent can folate supplementation reduce the incidence of neural tube defects? What caution is advised regarding folate intake and vitamin B12? What is fetal programming?

4. What is considered the most reliable indicator of an infant’s health? What are the two characteristics of a mother’s weight that can influence the infant’s birthweight? What is the greatest risk underweight women have during pregnancy? What medical complications can occur in an overweight woman during pregnancy? Cite the recommended weight gain for the following women during pregnancy: normal weight, underweight, overweight, and obese women. What contributes to weight gain during pregnancy? What is the average weight retained after a pregnancy?

5. What are the benefits of exercise during pregnancy? Cite the exercise guidelines during pregnancy. What is considered the ideal exercise during pregnancy? Why should pregnant women stay out of saunas, steam rooms, and hot tubs?

6. How many extra kcalories per day does a pregnant woman need? What is the RDA for protein during pregnancy? Why is it important to obtain essential fatty acids during pregnancy? What roles do folate and B_{12} play in a successful pregnancy? What nutrient is typically supplemented during the second and third trimesters for all pregnant women? Why is zinc supplementation not recommended during pregnancy? What bone-building nutrients are in great demand during pregnancy? What roles do vitamin D and calcium play during the early stages of pregnancy? Under what circumstance would an additional multivitamin-mineral supplement be recommended to pregnant women?

7. What seems to be responsible for early pregnancy nausea? What nutritional strategies will alleviate nausea? What nutritional strategies can prevent or alleviate constipation and/or hemorrhoids? How can heartburn be relieved? What is pica?

8. List the high-risk pregnancy factors. What is the effect of malnutrition on the following: fertility, early pregnancy, and fetal development? Define low birthweight. What complications are low-birthweight infants likely to experience? What percentage of infant deaths before one year of age are attributed to low birthweight?

9. What is gestational diabetes? How is gestational diabetes managed? What is gestational hypertension? How should edema be viewed during pregnancy? What are the warning signs of preeclampsia? What is eclampsia?

10. Of all the infants born, how many are born to teenage mothers? Cite the risks associated with teenage pregnancy. How much weight gain is recommended for young teenage mothers? Describe the common complications associated with pregnancy in women over thirty-five years old.

11. How does alcohol intake affect pregnancy? What are the risks associated with drug use and herbal supplements and fetal development? What risks are associated with smoking? What is sudden infant death syndrome? What are the recommendations for fish consumption during pregnancy and lactation? Explain how the following affect fetal development: environmental contaminants, megadoses of supplements, caffeine, dieting, and sugar substitutes.

VIDEO FOCUS POINTS

The following focus points are designed to help you get the most from the video segment of this lesson. Review them before watching the video. You may want to write notes to reinforce what you have learned.

1. What advice is given to pregnant women who want to begin an exercise program? Why is swimming considered the best exercise for pregnant women?

2. What is the primary weight gain attributed to during pregnancy? What advice is given regarding dieting during pregnancy?

3. What is the most at-risk age group for pregnancy? What is the greatest risk for teenage pregnancies? What is the recommendation for iron supplementation for pregnant women? What is the recommendation for folic acid supplementation for pregnant women? What condition will the intake of folic acid prevent?

4. Describe how being a single, working mother can impact personal fitness and nutrition goals.

Evaluating the Life Cycle

NAME: ______________________________ ID #: ________________________

The purpose of this exercise is to help you evaluate the human life cycle and the impact nutrition has on it.

DIRECTIONS:

1. Select one of the following life cycles.
2. Choose *at least* two topics under the life cycle.
3. Type (or write neatly) a 500- to 800-word paper elaborating on the topics.
4. Include your personal experience — positive or negative — related to the life cycle.
5. Cite two references. One reference may be your text.

PREGNANCY:

- The function of the placenta
- Critical periods of fetal development
- The effects of weight on pregnancy outcomes
- Exercise guidelines during pregnancy
- Energy and nutrient needs during pregnancy
- How malnutrition impacts pregnancy
- How diabetes and hypertension impact pregnancy
- Risks associated with adolescent pregnancy
- How drugs, alcohol, and smoking impact pregnancy outcomes

LACTATION AND INFANCY:

- Pros and cons associated with breastfeeding
- Pros and cons associated with formula feeding
- Energy and nutrient requirements of lactating women
- Energy and nutrient requirements of formula feeding women
- Energy and nutrient needs of the infant
- Differences between breast milk, formula, and cow's milk
- When solid foods should be introduced to the infant
- What solid foods should be introduced to the infant

—Continued on next page

CHILDHOOD AND ADOLESCENCE:

- Energy and nutrient requirements of children
- Impact of concentrated sweets on food selections of children
- The impact of malnutrition on growth and development of children
- How nutrition impacts behavior
- Energy and nutrient needs during adolescence
- How snacking impacts adolescent nutrition status
- How drugs, alcohol, and smoking impact adolescent nutrition

ADULTHOOD AND AGING:

- The effect of nutrition on longevity
- The effect of lifestyle choices on longevity
- The effect of heredity on longevity
- How aging affects nutrient intake
- Energy and nutrient requirements of older adults
- How exercise affects older adults
- Risk factors associated with malnutrition in older adults
- How nutrition affects conditions such as cataracts, arthritis, diabetes, and the brain
- How food choices and habits impact nutrition in the aging adult

PRACTICE TEST

The following items will help you check your understanding of this lesson. Compare your answers to the Answer Key at the end of the lesson. Review the course materials related to any incorrect answer.

Multiple Choice: Select the one choice that best answers the question.

1. What organ of the pregnant woman is central to the exchange of nutrients for waste products with the fetus?
   A. Uterus
   B. Vagina
   C. Placenta
   D. Amniotic sac

2. Studies report that folate supplements for women may lower the incidence of neural tube defects of infants when the vitamin is taken during the ________
   A. last trimester of pregnancy.
   B. second trimester of pregnancy.
   C. second and third trimesters of pregnancy.
   D. month before conception through the first trimester of pregnancy.

3. What is the recommended range of weight gain during pregnancy for a normal-weight woman?
   A. 10–18 pounds
   B. 19–24 pounds
   C. 25–35 pounds
   D. 35–40 pounds

4. To maintain physical fitness during pregnancy, all of the following activities are considered acceptable EXCEPT ________
   A. saunas.
   B. swimming.
   C. playing singles tennis.
   D. forty-five-minute balanced exercise sessions three times per week.

5. What is the recommended *increase* in energy intake for the second trimester of pregnancy?
   A. 200 kcalories per day
   B. 340 kcalories per day
   C. 450 kcalories per day
   D. 440 kcalories per day

6. Why is routine vitamin D supplementation during pregnancy NOT recommended?
   A. The RDA does not change.
   B. It may be toxic to the fetus.
   C. Self-synthesis rate of vitamin D increases markedly in pregnancy.
   D. Pregnancy leads to increased absorption efficiency of calcium, and therefore, extra vitamin D is not needed.

7. Which of the following is one of the recommendations to treat pregnancy-associated heartburn?
   A. Eat many small meals.
   B. Drink fluids only with meals.
   C. Exercise within thirty minutes after eating.
   D. Lie down within fifteen minutes after eating.

8. Which of the following is the standard classification for a low-birthweight infant?
   A. 3½ pounds or less
   B. 4 pounds or less
   C. 5½ pounds or less
   D. 6½ pounds or less

9. Which of the following is a characteristic of gestational diabetes?
   A. It predicts risk of diabetes for the infant.
   B. It occurs in over one-half of normal weight women.
   C. It leads to adult-onset diabetes in about a third of the women.
   D. It occurs more often in women with a history of having premature births.

10. Which of the following is a characteristic associated with adolescent pregnancy?
    A. The recommended weight gain is approximately 35 pounds.
    B. The incidence of stillbirths and preterm births is 5 to 10 percent lower compared with adult women.
    C. The incidence of pregnancy-induced hypertension is 5 to 10 percent lower compared with older women.
    D. The time in labor is usually shorter than for older women because there are fewer overweight teenagers.

11. Advice to pregnant women who want to exercise includes __________
    A. do not begin a new program.
    B. swimming, but not weight lifting.
    C. walking, but not too intensely.
    D. all of the above.

12. Dieting during pregnancy is __________
    A. recommended if the mother is overweight.
    B. not recommended at all.
    C. dependent upon the initial weight of the mother.
    D. directly correlated to high birthweight.

13. The recommendation for iron supplementation is __________
    A. greater for pregnant teens.
    B. about 30 to 60 milligrams per day.
    C. necessary to preserve iron stores.
    D. all of the above.

14. One of the reasons young mothers lose sight of personal fitness or nutrition goals is due to __________
    A. overwork and fatigue.
    B. stresses of being a new mother.
    C. postpartum depression.
    D. all of the above.

Matching: Select the letter next to the word or phrase that best matches the numbered statements. Answers are used only once.

_____15. A newly fertilized ovum

_____16. Fluid in which the fetus floats

_____17. Approximate weight, in pounds, of an average newborn baby

_____18. Number of grams of extra protein per day recommended for a pregnant woman

_____19. A recommended practice to prevent or relieve heartburn

A. Six
B. Seven
C. Twenty-five
D. Zygote
E. Amniotic
F. Take antacids
G. Eat small, frequent meals

Essay: Reflect on the item(s) listed below; one or more may be included on the exam.

20. What are several benefits of exercise specifically for the pregnant woman? What types of exercise should be avoided and why?

21. Describe the risks associated with adolescent pregnancy.

22. Discuss the consequences of maternal alcohol intake on fetal development.

ANSWER KEY

The following provides the answers and references for the Practice Test questions.

	Answer	Learning Objective	Reference
1.	C	LO 1	textbook, p. 470
2.	D	LO 3, 6	textbook, p. 473
3.	C	LO 4	textbook, p. 477
4.	A	LO 5	textbook, pp. 478–479
5.	B	LO 6	textbook, p. 480
6.	A	LO 6	textbook, p. 482
7.	A	LO 7	textbook, p. 484
8.	C	LO 9	textbook, p. 484
9.	C	LO 10	textbook, p. 487
10.	A	LO 11	textbook, p. 488
11.	D	LO 5	Video
12.	B	LO 8	Video
13.	D	LO 6	Video
14.	D	LO 13	Video
15.	D	LO 2	textbook, p. 471
16.	E	LO 2	textbook, p. 470
17.	B	LO 9	textbook, p. 478
18.	C	LO 6	textbook, p. 480
19.	G	LO 7	textbook, p. 484
20.		LO 5, 13	textbook, pp. 478–479
21.		LO 11	textbook, p. 488
22.		LO 12	textbook, pp. 501–503

Lesson 19

Life Cycle: Lactation and Infancy

LESSON ASSIGNMENTS

Text: Whitney and Rolfes, *Understanding Nutrition*, Chapter 15, "Life Cycle Nutrition: Pregnancy and Lactation," pp. 492–497. Chapter 16, "Life Cycle Nutrition: Infancy, Childhood, and Adolescence," pp. 505–516

Video: "Life Cycle: Lactation and Infancy" from the series *Nutrition Pathways*

Optional Web Activities:
Consult your instructor and/or syllabus for any assigned activities.

OVERVIEW

You have learned how nutrition impacts the health of a pregnant woman and her fetus; now let's examine the role nutrition plays when a woman prepares to breastfeed her infant. In this lesson you will consider the advantages and disadvantages of both breastfeeding and infant formula and come to understand why women select the method of feeding that is best for them and their infants. You will discover circumstances which may prevent a mother from breastfeeding, learn why and when cow's milk should be introduced to the infant, and learn when to introduce solid foods and what are the best foods to introduce to the infant.

LEARNING OBJECTIVES

Upon completing this lesson, you should be able to:

1. Describe the positive side effects associated with breastfeeding for the infant and for the mother.
2. Cite the kcalorie requirements of the lactating woman.
3. Explain how lactation affects weight loss after pregnancy.

4. Explain how exercise affects lactation.

5. Describe the nutrient needs of the breastfeeding and formula-feeding woman.

6. Identify substances and conditions that may prohibit breastfeeding.

7. Describe the energy and nutrient needs of the infant.

8. Explain the differences between breast milk, infant formula, and cow's milk.

9. Describe the process of introducing solid foods to the infant.

10. Explain the effect of stress on breastfeeding.

TEXT FOCUS POINTS

The following focus points are designed to help you get the most from your reading. Review them, then read the assignment. You may want to write notes to reinforce what you have learned.

1. What protection is provided to an infant who drinks mother's milk? What physiological and other benefits are provided to a woman who breastfeeds? Define *prolactin* and *oxytocin*.

2. In order to produce milk, how many extra kcalories per day must a lactating woman take in during the first six months of lactation? What is the recommendation in kcalories per day for the lactating mother to meet this energy need?

3. What is the effect of severe kcalorie restriction on breast milk? How much weight per month do most women lose during the first four to six months of lactation?

4. How does intense exercise affect lactation or breast milk production?

5. What is the effect of nutrient deficiencies on the quantity and quality of breast milk? What is the recommendation for fluids during lactation? Why might iron supplements be necessary for lactating women? What foods might affect the flavor of breast milk?

6. Describe the effects of the following on breast milk production and/or infant development: alcohol, drugs, caffeine, smoking, environmental contaminants, viruses, and diabetes.

7. How many kcalories per day do infants require based on body weight? On what is the RDA for protein based for infants? On what are vitamin and mineral recommendations based for infants? Why can water be so easily lost from an infant's body?

8. What is the carbohydrate found in breast milk, and what is its function? What is the chief protein found in breast milk? What vitamin is typically deficient in breast milk? What supplements are necessary for breastfed infants? What is colostrum? What are bifidus factors? What is lactoferrin?

9. Compare the nutrients of breast milk, with infant formula for iron and protein content. What causes "nursing bottle tooth decay"? Describe the special needs of preterm infants. Why is cow's milk inappropriate during the first year?

10. At what age should solid foods be introduced to the infant? What are the indications that solid foods may be given to an infant? Why should foods be introduced one at a time? What solid food is the first one generally introduced to an infant and why? What grain is typically introduced last and why? At what age does the infant begin to require more iron, and what food sources will provide adequate iron? Why do some experts recommend introducing vegetables before fruit to infants? What foods should be omitted in an infant's diet and why? What is botulism?

VIDEO FOCUS POINTS

The following focus points are designed to help you get the most from the video segment of this lesson. Review them before watching the video. You may want to write notes to reinforce what you have learned.

1. Compare the energy needs of a breastfed infant to a formula-fed infant.

2. What are the guidelines that indicate a breastfed infant is getting enough food and kcalories?

3. What are the nutritional requirements for a woman who is not breastfeeding?

4. What is the recommended age to introduce cow's milk? At what age should breastfed infants begin to receive iron supplementation?

5. What is the recommendation regarding introducing fruit versus vegetables?

6. What is the "letdown reflex"? How does stress affect the letdown reflex?

PRACTICE TEST

The following items will help you check your understanding of this lesson. Compare your answers to the Answer Key at the end of the lesson. Review the course materials related to any incorrect answer.

Multiple Choice: Select the one choice that best answers the question.

1. Which of the following describes the findings from studies of lactating women who exercised intensely compared with sedentary, lactating women?
   A. They had similar energy intakes.
   B. Their milk was more nutrient-dense.
   C. They had a slightly greater amount of body fat.
   D. Their milk contained more lactic acid which alters taste.

2. Which of the following reflects one of the features of alcohol intake on lactation?
   A. It does not pass into the milk.
   B. It mildly stimulates milk production.
   C. It hinders the infant's ability to breastfeed.
   D. It passes into the milk but is degraded by enzymes in breast tissue.

3. Under which of the following circumstances would it still be acceptable for a mother to breastfeed?
   A. She has alcohol abuse.
   B. She has a drug addiction.
   C. She has an ordinary cold.
   D. She has a communicable disease.

4. What is the chief protein in human breast milk?
   A. Casein
   B. Lactose
   C. Albumin
   D. Alpha-lactalbumin

5. Which of the following should NOT be used to feed an infant?
   A. Whole milk
   B. Ready-to-feed formula
   C. Liquid concentrate formula appropriately diluted
   D. Powdered formula or evaporated milk formula appropriately reconstituted

6. Which of the following represents a good age to introduce solid foods to infants?
   A. Two weeks
   B. Two months
   C. Five months
   D. One year

7. Low-fat or nonfat milk should not be given routinely to a child until after the age of __________
   A. two weeks.
   B. three months.
   C. two years.
   D. six years.

8. Compared to formula-fed infants, breastfed infants seem to need __________
   A. less energy.
   B. more energy.
   C. more iron.
   D. less iron.

9. One of the ways you can tell if an infant is getting enough food is __________
   A. by the number of wet diapers per day.
   B. if the infant is gaining two pounds a week.
   C. if the infant sleeps through the night at three weeks.
   D. by the number of feedings per day.

10. A woman who is not breastfeeding __________
    A. has the same nutrient requirements as nonlactating women.
    B. needs more nutrients than nonlactating women.
    C. has the same nutrient requirements as lactating women.
    D. has a greater need for water than lactating women.

11. Nutrition scientists recommend that cow's milk be given to infants __________
    A. as soon as they can sit up.
    B. after the age of twelve months.
    C. who are eating solid foods.
    D. any of the above.

12. Some research shows that introducing fruits before vegetables __________
    A. will cause the infant to develop a sweet tooth.
    B. causes more allergic reactions.
    C. does not really matter; you can introduce either one first.
    D. causes excess weight gain in the infant.

13. The order of introduction of fruits and vegetables to infants __________
    A. does not seem to matter.
    B. should be to start with bananas.
    C. should be to start with squash.
    D. should be to mix them together at the same meal.

14. Stress in the mother who breastfeeds has been shown to __________
    A. inhibit the quality of milk produced.
    B. cause greater milk production.
    C. inhibit the letdown reflex.
    D. cause weight loss in the infant.

Fill in the Blank: Insert the correct word or words in the blank for each item.

15. At least one criteria that indicates whether an infant is getting enough food is ______________________________.

Matching: Select the letter next to the word or phrase that best matches the numbered statements. Answers are used only once.

_____16. Typical daily energy need, in kcalories per kilogram of body weight of an infant

_____17. Essential fatty acid in breast milk

_____18. Process whereby breast milk is gradually replaced by formula or semisolid foods

A. 100
B. 120
C. Linoleic
D. Weaning
E. Vitamin B
F. Vitamin C

Essay: Reflect on the item(s) listed below; one or more may be included on the exam.

19. Compare and contrast major differences in nutrient content of breast milk and cow's milk.

20. Explain the energy needs for breastfeeding in light of the mother's desire to lose the extra weight from pregnancy.

ANSWER KEY

The following provides the answers and references for the Practice Test questions.

	Answer	Learning Objective	Reference
1.	D	LO 4	textbook, p. 494
2.	C	LO 6	textbook, p. 496
3.	C	LO 6	textbook, pp. 496–497
4.	D	LO 8	textbook, p. 509
5.	A	LO 8	textbook, p. 509
6.	C	LO 9	textbook, pp. 512–513
7.	C	LO 7	textbook, pp. 512–513
8.	A	LO 1	Video
9.	A	LO 7	Video
10.	A	LO 2, 5	Video
11.	B	LO 8	Video
12.	C	LO 9	Video
13.	A	LO 9	Video
14.	C	LO 10	Video
15.	May include the following: six to eight wet diapers per day one bowel movement per day happy and alert baby appropriate weight gain	LO 7	Video
16.	A	LO 7	textbook, p. 506
17.	C	LO 8	textbook, p. 509
18.	D	LO 9	textbook, p. 511
19.		LO 8	textbook, pp. 511–512
20.		LO 3	textbook, pp. 493–494

Lesson 20

Life Cycle: Childhood and Adolescence

LESSON ASSIGNMENTS

Text: Whitney and Rolfes, *Understanding Nutrition*, Chapter 16, "Life Cycle Nutrition: Infancy, Childhood, and Adolescence," pp. 516–548

Video: "Life Cycle: Childhood and Adolescence" from the series *Nutrition Pathways*

Optional Web Activities:
Consult your instructor and/or syllabus for any assigned activities.

OVERVIEW

Childhood and adolescence are two periods during the life cycle that have their own unique nutrition challenges. In this lesson, you will consider the changes that take place during childhood and adolescence and how nutrition impacts those changes. You will begin to see how a child's appetite and nutrient needs reflect the different stages of growth throughout childhood; how poor nutrition impacts growth, learning, and behavior; and how parents can influence children's eating habits and lifestyles for the rest of their lives. You will learn that nutrient requirements rise dramatically during adolescence along with rapid growth spurts, and that teenagers have certain challenges they did not face as children, such as the strong influence of peers on lifestyle and nutrition habits. Finally, you will examine ways parents can help teenagers make wise choices about nutrition and lifestyles even when parents are no longer the focal point of their adolescents' lives.

LEARNING OBJECTIVES

Upon completing this lesson, you should be able to:

1. Cite feeding guidelines for toddlers.
2. In general terms, describe the kcalorie and nutrient requirements for children.
3. Explain the difference between food allergies and food intolerance and how they impact child growth and development.
4. Describe how malnutrition impacts growth and development.
5. Explain how nutrition impacts behavior.
6. State the strategy parents and schools can use to impact future nutrition behavior in children.
7. Describe the impact of childhood obesity on growth, physical health, and psychological problems.
8. In general terms, describe the kcalorie and nutrient needs of adolescents.
9. Explain how snacking and fast foods impact adolescent nutrition.
10. Describe how drug use impacts the nutrition status of the adolescent.
11. Explain how the media influence adolescent behavior with regard to body image.

TEXT FOCUS POINTS

The following focus points are designed to help you get the most from your reading. Review them, then read the assignment. You may want to write notes to reinforce what you have learned.

1. List helpful feeding guidelines aimed at toddlers. How many kcalories does a one-year-old need? A six-year-old? A ten-year-old? How do protein and other nutrient needs change as a child grows?
2. What is the most common nutrient deficiency in children in the United States and Canada? How much iron must a child consume to prevent iron deficiency?

Explain how MyPlate can be utilized when planning children's meals. What percent of children eat the suggested servings from the Food Patterns for children? What nutrients typically fall below recommendations in children?

3. What are the effects on a child who does not eat a nutritious breakfast? What can teachers do to prevent a midmorning slump in children? What are the immediate and future effects of iron deficiency? What are general symptoms of a child suffering from nutrient deficiencies? Explain the connection between malnutrition in children and lead poisoning.

4. What is hyperactivity? How can hyperactivity be controlled? How does sugar intake affect children with hyperactivity? What is the difference between food intolerance and food allergies? What three foods cause the majority of allergic reactions? What percent of food allergies in children can be attributed to peanuts, eggs, and milk?

5. What factor predicts an early increase in children's BMI and increases the chance of becoming obese? What two factors are responsible for the fact that children are much heavier today than twenty years ago? What is the single most important problem facing obese children? How does obesity affect growth, physical health, and psychological development in children? What dietary strategies can be used with children in order to prevent or treat obesity? What is the most important thing a parent can do to encourage children to become more physically active?

6. Why should parents encourage children to help plan and prepare meals? What should parents do when a child won't accept new foods? When should new foods be introduced? What should parents do regarding snacking? What is the single most important influence on a child's eating habits?

7. At what ages do males and females go through adolescent growth spurts? How much weight do male and female adolescents gain during the growth spurt? Why do boys require more kcalories per day than do girls of the same age? Why do both sexes require more iron and calcium at the onset of adolescence?

8. What fraction of a teen's daily food intake is typically from snacks? What can parents do to ensure healthful snacking? What nutrients are generally lacking from fast foods that need to be available in other meals? What are the consequences of frequent intake of soft drinks?

9. What is the effect of marijuana on appetite and body weight? What is the effect of cocaine on appetite and body weight? What is the general effect of alcohol on nutrition status? How does smoking affect health, appetite, and weight? What nutrients do smokers typically take in less of than nonsmokers do?

VIDEO FOCUS POINTS

The following focus points are designed to help you get the most from the video segment of this lesson. Review them before watching the video. You may want to write notes to reinforce what you have learned.

1. What is the first concrete element that passes between a primary caregiver and an infant? How does the importance of food impact future eating behavior? What can happen to children when parents do not set limits regarding food intake?

2. How can parents teach children about nutrition and encourage responsibility toward food?

3. What impact can parents' habitual overeating have on children?

4. How do programs, such as Shape Down, help adolescents lose weight? What factor(s) most influence weight changes in adolescents?

5. How do media expectations regarding body image impact self-image and behavior in adolescent girls?

6. How can food choices as an adolescent influence future development of type 2 diabetes?

PRACTICE TEST

The following items will help you check your understanding of this lesson. Compare your answers to the Answer Key at the end of the lesson. Review the course materials related to any incorrect answer.

Multiple Choice: Select the one choice that best answers the question.

1. What should be the parent's response when a one-year-old child wants to clumsily spoon-feed himself?
   A. Punish the child.
   B. Let the child eat with his fingers instead.
   C. Let the child try to feed himself so that he will learn.
   D. Gently take the spoon back and feed the child with it.

2. The consumption of milk by children should not exceed four cups per day in order to lower the risk for __________
   A. solute overload.
   B. iron deficiency.
   C. vitamin A toxicity.
   D. vitamin D toxicity.

3. Which of the following is a feature of iron nutrition in the very young?
   A. Iron deficiency is most prevalent in children aged two to three years old.
   B. The supply of stored iron becomes depleted after the birthweight triples.
   C. Infants with iron-deficiency anemia demonstrate abnormal motor development.
   D. Serum ferritin concentrations fall in infants who start drinking whole milk at three months of age but not at six months of age.

4. An adverse reaction to food that does NOT signal the body to form antibodies is termed a _________
   A. food allergy.
   B. food intolerance.
   C. mild food challenge.
   D. transient food episode.

5. Even in preschoolers whose habits are being established, existing dietary attitudes are relatively resistant to change. How should wise parents react?
   A. Be patient and persistent.
   B. Impose their own eating habits on the children.
   C. Wait until the children start school to initiate changes.
   D. Exert continuous pressure to initiate good food habits.

6. The adolescent growth spurt __________
   A. affects the brain primarily.
   B. decreases total nutrient needs.
   C. affects every organ except the brain.
   D. begins and ends earlier in girls than in boys.

7. Approximately what fraction of an average teenager's daily energy intake is derived from snacks?
   A. One-fourth
   B. One-third
   C. One-half
   D. Two-thirds

8. When food is overly important or not attended to properly in the family, experts have found that there __________
   A. are more eating disorders.
   B. are fewer incidences of binging and purging.
   C. is a greater chance for discipline problems.
   D. all of the above.

9. Parents can teach children to become responsible with regard to food intake by __________
   A. modeling the behavior they want the children to learn.
   B. creating positive learning experiences around food.
   C. giving children a certain amount of authority with regard to food choices.
   D. all of the above.

10. Children will learn to disregard their own internal hunger and satiety cues if they ________
    A. observe their parents habitually overeating.
    B. watch too much television, especially food advertisements.
    C. eat too many complex carbohydrates during the day.
    D. drink too many soft drinks.

11. The weight loss program, Shape Down, helps adolescents lose excess weight by ________
    A. treating not only the body but also the mind.
    B. utilizing an interdisciplinary team of professionals.
    C. involving all family members in the program.
    D. all of the above.

12. Adolescent girls who do not measure up to media portrayal of the perfect young body image ________
    A. may develop eating disorders.
    B. develop a stronger self-image and more confidence.
    C. usually are more accepting of their bodies in spite of the magazines.
    D. frequently commit suicide.

13. Adolescents who snack on high-fat foods or sugary foods will develop type 2 diabetes as adults if they ________
    A. are genetically predisposed.
    B. do not change their eating habits.
    C. gain too much weight.
    D. all of the above.

Matching: Select the letter next to the word or phrase that best matches the numbered statements. Answers are used only once.

_____ 14. Approximate percentage of young children with food allergies

_____ 15. Period in life when an individual becomes physically capable of reproduction

_____ 16. Substance known to suppress appetite and increase rate of energy expenditure

A. Six
B. Ten
C. Nicotine
D. Puberty
E. Adulthood

Essay: Reflect on the item(s) listed below; one or more may be included on the exam.

17. Give examples of how hunger and nutrient deficiencies affect behavior in children.

18. List six nutrition problems associated with drug abuse and tobacco use in adolescents.

ANSWER KEY

The following provides the answers and references for the Practice Test questions.

	Answer	Learning Objective	Reference
1.	C	LO 1	textbook, p. 516
2.	B	LO 4	textbook, p. 516
3.	C	LO 2	textbook, p. 521
4.	B	LO 3	textbook, p. 526
5.	A	LO 6	textbook, pp. 531–532
6.	D	LO 8	textbook, p. 537
7.	A	LO 9	textbook, p. 539
8.	A	LO 6, 7	Video
9.	D	LO 6	Video
10.	A	LO 7	Video
11.	D	LO 7	Video
12.	A	LO 7, 11	Video
13.	A	LO 9	Video
14.	A	LO 3	textbook, pp. 524–525
15.	D	LO 8	textbook, p. 537
16.	C	LO 10	textbook, p. 548
17.		LO 5	textbook, pp. 520–521
18.		LO 10	textbook, p. 548

Lesson 21

Life Cycle: Adulthood and Aging

LESSON ASSIGNMENTS

Text: Whitney and Rolfes, *Understanding Nutrition*, Chapter 17, "Life Cycle Nutrition: Adulthood and the Later Years," pp. 551–578

Video: "Life Cycle: Adulthood and Aging" from the series *Nutrition Pathways*

Optional Web Activities:
Consult your instructor and/or syllabus for any assigned activities.

OVERVIEW

Wise food choices made throughout the growth years and into adulthood can help a person meet the many physical, emotional, and mental challenges of day-to-day living. Appropriate food choices and lifestyle habits can also help us to achieve freedom from diseases such as cardiovascular disease, diabetes, and cancer. The focus of this lesson reflects how nutrition and lifestyle habits support or weaken a person's quest for longevity and wellness throughout adulthood and into the later years. In addition, you will see how genetics and the aging process play important roles in the quality of one's life; what nutrients are needed as people grow older; and how older adults can solve problems associated with a limited budget and nutrition concerns.

LEARNING OBJECTIVES

Upon completing this lesson, you should be able to:

1. Describe how nutrition affects longevity.
2. Describe how lifestyle affects longevity.
3. Explain the changes that can affect aging or nutrient intake.
4. Cite the nutrient and energy needs of older adults.
5. Explain how exercise affects older adults.
6. Cite risk factors associated with malnutrition in older adults.
7. State how nutrition may affect cataracts, arthritis, diabetes, and the aging brain.
8. Cite strategies for growing old gracefully.
9. Describe how food choices and eating habits impact nutrition in the elderly.

TEXT FOCUS POINTS

The following focus points are designed to help you get the most from your reading. Review them, then read the assignment. You may want to write notes to reinforce what you have learned.

1. What is the fastest growing age group in the United States? What are the life expectancies for women and men in the United States? What is the potential life span of humans? List six lifestyle behaviors that have the greatest influence not only on health but also on physiological age.

2. What are the exercise guidelines for older adults? What are the benefits of exercise for older adults? What is the most powerful predictor of a person's mobility in later years? What is the effect of strength training on aging? Why is general fear in the elderly associated with falls and how can regular physical activity alleviate that fear? What is a recommended exercise regimen for an older, healthy person who is just beginning an exercise program?

3. What is the effect of moderate kcalorie restriction on humans? What are the recommendations regarding body weight in older adults? What happens to body

composition as we age? What happens to the immune system as we age? What happens to the gastrointestinal (GI) tract as we age? What is atrophic gastritis? How do dental problems affect the elderly? What happens to the senses of taste and smell in the elderly? What other changes may affect nutrient status of the elderly?

4. Why is dehydration a problem for many older people? At what age does the RDA for energy begin to decrease? Explain how the need for protein, carbohydrate, fiber, and fat changes as we age.

5. Why are many elderly people deficient in vitamin B_{12}? What vitamin is absorbed and stored better in the elderly? What vitamin poses a potential risk for deficiency in the elderly and why? Generally speaking, what are the recommendations regarding calcium intake in older people? How does iron deficiency in older adults compare to younger people? What might affect an older person's iron status? What is the common zinc status of elderly people and what are symptoms of deficiency?

6. Under what circumstances might older adults need supplementation?

7. What are cataracts? What percent of people over age sixty-five have cataracts? What nutrients are associated with cataract development? Define *osteoarthritis* and *rheumatoid arthritis*. How does nutrition affect arthritis? What are considered normal changes in the brain as people age? What is senile dementia? What is Alzheimer's Disease? What might be the most important nutrition concern for people with Alzheimer's Disease? How does nutrition impact the aging brain?

8. What strategies can people employ to help them age healthfully? What are the risk factors for malnutrition in older adults?

9. What are the typical food preferences in the elderly? Cite suggestions that can help older single people solve problems associated with buying, storing, and preparing foods for one person.

VIDEO FOCUS POINTS

The following focus points are designed to help you get the most from the video segment of this lesson. Review them before watching the video. You may want to write notes to reinforce what you have learned.

1. How does aging affect the need for high energy foods?
2. What is the basic idea scientists are trying to get across to people with regard to aging?
3. What type of exercise is recommended to prevent loss of muscle mass? What type of exercise is recommended to prevent the onset of osteoporosis?
4. What are the effects of alcohol on an older person?
5. How does diabetes impact the aging process?

PRACTICE TEST

The following items will help you check your understanding of this lesson. Compare your answers to the Answer Key at the end of the lesson. Review the course materials related to any incorrect answer.

Multiple Choice: Select the one choice that best answers the question.

1. Studies of adults show that longevity is related, in part, to all of the following EXCEPT ________
   A. weight control.
   B. regularity of meals.
   C. short periods of sleep.
   D. no or moderate alcohol intake.

2. Which of the following is a finding from studies of diet restriction in rats?
   A. Restriction of specific nutrients exerted antiaging effects.
   B. Energy-restricted diets led to life extension in 90 percent of the rats.
   C. Energy-restricted diets led to lowering of the metabolic rate and body temperature.
   D. Restriction of food intake only after rats reached maturity, but not before, resulted in extension of life span.

3. Atrophic gastritis is typically characterized by all of the following signs EXCEPT __________
   A. inflamed stomach mucosa.
   B. lack of hydrochloric acid.
   C. abundant bacteria in the stomach.
   D. insufficient secretion of pepsinogen and gastrin.

4. A condition that increases the likelihood of iron deficiency in older people is __________
   A. lack of intrinsic factor.
   B. loss of iron due to more frequent running activity.
   C. blood loss from yearly physical testing procedures.
   D. poor iron absorption due to reduced stomach acid secretion and/or use of antacids.

5. What nutrients appear to protect against cataract formation?
   A. Iron and calcium
   B. Chromium and zinc
   C. Vitamin B_{12} and folate
   D. Vitamin C and vitamin E

6. What organ is affected by macular degeneration?
   A. Bone
   B. Eyes
   C. Liver
   D. Kidneys

7. Which of the following is a characteristic of Alzheimer's disease?
   A. It affects 60 percent of those over eighty years of age.
   B. It is responsive to dietary choline supplementation.
   C. It is associated with stability of brain nerve cell number.
   D. It is associated with clumps of beta-amyloid protein in the brain.

8. If scientists had their way, when it was time to die, we would all __________
   A. go, but not fight all the way.
   B. go quickly, like a light bulb.
   C. be glad that medical science kept us going so long.
   D. none of the above.

9. Alcohol can affect an older adult ________
   A. by increasing blood alcohol level quickly.
   B. because they are more sensitive to alcohol's effects.
   C. because they have less body water to dilute it.
   D. all of the above.

10. A person who has lived with diabetes for twenty to forty years __________
   A. is more likely to have complications associated with the disease.
   B. is not likely to suffer any more ill effects than a younger person with the disease.
   C. should pay especially close attention to the level of glycemic control.
   D. A and C.

Fill in the Blank: Insert the correct word or words in the blank for each item.

11. As you age, there is less of a need for foods that are ________________.

12. As a person ages, there is a relative loss of ____________________.

Matching: Select the letter next to the word or phrase that best matches the numbered statements. Answers are used only once.

_____ 13. Dietary restriction of this extends lifespan

_____ 14. Water intake recommendation for adults in ounces per kilogram of body weight

_____ 15. Percentage of elderly population who are deficient in vitamin B_{12}

A. One
B. Fifteen
C. Aspirin
D. Energy
E. Vitamin A

Essay: Reflect on the item(s) listed below; one or more may be included on the exam.

16. Discuss the roles of fitness and stress on longevity.

17. List the factors that increase the risk for vitamin B_{12} and iron deficiency in older adults.

ANSWER KEY

The following provides the answers and references for the Practice Test questions.

	Answer	Learning Objective	Reference
1.	C	LO 1, 2	textbook, p. 553
2.	C	LO 1	textbook, pp. 554–555
3.	D	LO 3	textbook, p. 562
4.	D	LO 3	textbook, p. 563
5.	D	LO 7	textbook, p. 564
6.	B	LO 6, 7	textbook, p. 564
7.	D	LO 7	textbook, pp. 566–567
8.	B	LO 8, 9	Video
9.	D	LO 9	Video
10.	D	LO 8, 9	Video
11.	High in kcalories (fat, energy)	LO 3	Video
12.	Muscle mass (lean body mass, muscle strength)	LO 4	Video
13.	D	LO 4	textbook, pp. 560–561
14.	A	LO 4	textbook, p. 560
15.	B	LO 6	textbook, p. 562
16.		LO 1, 2, 5	textbook, pp. 553–556
17.		LO 6	textbook, pp. 562–563

Lesson 22

Diet and Health: Cardiovascular Disease

LESSON ASSIGNMENTS

Text: Whitney and Rolfes, *Understanding Nutrition*, Chapter 18, "Diet and Health," pp. 585–597

Video: "Diet and Health: Cardiovascular Disease" from the series *Nutrition Pathways*

Related Activity:

This activity is not required unless your instructor assigns it. It is offered as a suggestion to help you learn more about the material presented in this lesson. Refer to your syllabus to determine whether this activity has been assigned.

Complete the attached form for the following related activity, and return it to your instructor according to the established deadline.

❒ "Assessing the Risk for Disease"

Optional Web Activities:

Consult your instructor and/or syllabus for any assigned activities.

OVERVIEW

In this lesson you will learn about the disease that kills more people each year than any other disease—cardiovascular disease (CVD). Under the umbrella of CVD terms are *atherosclerosis*, *heart disease*, *hypertension*, and *stroke*. You will learn the ways each develops, the risk factors associated with each type, and diet and lifestyle choices that can protect you from developing CVD or help you recover if you already have cardiovascular disease.

LEARNING OBJECTIVES

Upon completing this lesson, you should be able to:

1. Identify risk factors for chronic disease.
2. Define the terms associated with cardiovascular disease (CVD).
3. Explain how atherosclerosis develops.
4. Describe the risk factors associated with CVD.
5. Present the recommendations for reducing CVD.
6. Explain how hypertension develops.
7. Describe the risk factors associated with hypertension.
8. Cite the recommendations for reducing hypertension.

TEXT FOCUS POINTS

The following focus points are designed to help you get the most from your reading. Review them, then read the assignment. You may want to write notes to reinforce what you have learned.

1. List the risk factors for chronic disease that can and cannot be changed. Define the following terms: *cardiovascular disease* (CVD); *coronary heart disease* (CHD); *atherosclerosis; plaque*; *prehypertension*; *hypertension*. What are the leading causes of CVD death for adults?

2. How does atherosclerosis usually begin? Define c-reactive protein. How is c-reactive protein used to assess risk for a heart event? By what age do most people have well-developed plaques? What are blood platelets, and how do they impact blood clotting? What compounds regulate the action of platelets? What type of fatty acid produces eicosanoids that favor heart health? How does plaque affect the blood pressure? Define the following terms: *coronary thrombosis*, *cerebral thrombosis*, *embolus*, *heart attack*, and *stroke*.

3. Describe in detail the major risk factors associated with heart disease. List the five risk factors that can be modified by diet or weight loss. What is considered

a desirable blood cholesterol level? What risk factors cannot be modified? What two risk factors may be the most influential on CHD? What three therapies are always recommended before drugs to treat CVD?

4. What are the two recommendations to reduce CVD risk? What major changes may be included in the treatment of CVD? On what do dietary recommendations to reduce risk for CHD focus? How do triglycerides impact heart disease? What are recommended levels for triglycerides? For LDL, HDL, and total cholesterol? What advice is offered to lower LDL? What advice is given by the American Heart Association (AHA) regarding energy intake? How does weight loss impact heart disease risk? What recommendation is given regarding total fat, saturated fat, and cholesterol intakes? What effect might monounsaturated fat, omega-3 fatty acids, and *trans*-fatty acids have on CHD risk? What are the recommendations and benefits of soluble fiber? How does soy protein affect LDL cholesterol? How does alcohol intake affect HDL cholesterol? How does exercise impact CHD risk and cholesterol levels? How do diet and exercise affect HDL? If diet and exercise therapy fail, what is the recommended treatment for atherosclerosis?

5. What percent of the adult population has hypertension? Briefly describe how blood pressure is regulated. How does obesity affect blood pressure? How does salt impact blood pressure? How does salt impact blood pressure?

6. Describe the risk factors associated with hypertension. Which of the risk factors for hypertension might be modified by diet?

7. What is the single most effective step people can take against hypertension? What is considered normal blood pressure for adults? What blood pressure reading indicates hypertension? How does body fat distribution affect blood pressure? How does physical activity impact blood pressure? How does alcohol intake impact hypertension? What does research state regarding salt and/or sodium intake and high blood pressure? Name other minerals that play a role in the prevention of hypertension. Briefly describe the Dietary Approaches to Stop Hypertension (DASH) eating plan. If diet and physical activity fail to reduce blood pressure, what is the recommended drug therapy?

VIDEO FOCUS POINTS

The following focus points are designed to help you get the most from the video segment of this lesson. Review them before watching the video. You may want to write notes to reinforce what you have learned.

1. What is the most common form of cardiovascular disease and how does it develop? What are the controllable risk factors for CVD? What are the risk factors that can not be controlled? What causes hypertension? What is the recommended blood pressure? What causes a stroke? When do women typically develop heart disease compared to men? What are the recommended levels of cholesterol? How can cholesterol levels be improved? How does being overweight impact heart disease?

2. How does exercise impact heart disease? What is the DASH diet dietary approach to stopping hypertension? How do sodium and dietary fat intake affect heart disease?

Assessing the Risk for Disease

NAME: ______________________________ ID #: ________________________

This assignment is designed to help you examine nutrition and lifestyle choices as they relate to risks associated with disease.

DIRECTIONS:

1. Select one of the following lessons pertaining to diet and health.
2. Provide details based on your personal experience, or interview someone you know (identities may be kept anonymous) who has (or had) CVD, cancer, HIV/AIDS, or diabetes.
3. Respond to the questions below.
4. Type (or print neatly) the questions followed by the answers.
5. Cite two references. One reference may be your text.

CARDIOVASCULAR DISEASE (CVD):

- Do you (or someone you know) have CVD?
- Does (did) that person have heart disease, hypertension, or stroke?
- When was it diagnosed?
- Describe the symptoms displayed by the person prior to diagnosis.
- Was that person aware of the risk factors associated with the development of CVD prior to diagnosis?
- What were the nutrition habits of the person prior to diagnosis of the disease?
- Describe the lifestyle habits of the person prior to diagnosis with regard to the following: exercise, weight control, smoking, alcohol intake, and stress.
- How did heredity contribute to the disease?
- Describe the treatment for the condition with regard to medication, surgery, nutrition intervention, and exercise intervention.
- Upon being diagnosed, describe how the person changed with regard to the following: nutrition, exercise, other lifestyle habits.
- Describe your personal risk for CVD.
- If the person with the disease is a family member, explain how this experience has affected you with regard to your lifestyle choices.
- What three pieces of advice would you give to someone in a similar situation to help that person reduce the risk for CVD or to treat the disease?

—Continued on the back

CANCER, IMMUNE DEFICIENCY, OR AIDS:

- Do you (or someone you know) have cancer, an immune deficiency, or HIV/AIDS?
- If the person has (had) cancer (HIV/AIDS), what type of cancer is (was) it?
- When was it diagnosed?
- Describe the symptoms displayed by the person prior to diagnosis.
- Was that person aware of the risk factors associated with the development of cancer (HIV/AIDS)?
- What were the nutrition habits of the person prior to diagnosis of the disease?
- Describe the lifestyle habits of the person prior to diagnosis with regard to the following: exercise, weight control, smoking, alcohol intake, stress.
- How did heredity contribute to the disease?
- Describe the treatment for the condition with regard to medication, surgery, nutrition intervention, and exercise intervention.
- Upon being diagnosed, describe how the person changed with regard to the following: nutrition, exercise, other lifestyle habits.
- Describe your personal risk for cancer or HIV/AIDS.
- If the person with the disease is a family member, explain how this experience has affected you with regard to your lifestyle choices.
- What three pieces of advice would you give to someone in a similar situation to help that person reduce the risk for cancer (HIV/AIDS) or to treat the disease?

DIABETES:

- Do you (or someone you know) have diabetes?
- Is it type 1 or type 2 diabetes?
- When was the person diagnosed?
- Describe the symptoms displayed by the person prior to diagnosis.
- Was the person aware of the risk factors associated with the development of diabetes?
- What were the nutrition habits of the person prior to diagnosis?
- Describe the lifestyle habits of the person prior to diagnosis with regard to the following: exercise, weight control, smoking, alcohol intake, stress.
- How did heredity contribute to the disease?

—*Continued on next page*

- Describe the treatment for diabetes with regard to medication, surgery, nutrition intervention, and exercise intervention.
- Upon being diagnosed, describe how the person changed with regard to the following: nutrition, exercise, other lifestyle habits.
- Describe your personal risk for diabetes.
- If the person with the disease is a family member, explain how this experience has affected you with regard to your lifestyle choices.
- What three pieces of advice would you give to someone in a similar situation to help that person reduce the risk for diabetes or treat the disease?

PRACTICE TEST

The following items will help you check your understanding of this lesson. Compare your answers to the Answer Key at the end of the lesson. Review the course materials related to any incorrect answer.

Multiple Choice: Select the one choice that best answers the question.

1. What is the term given to mounds of lipid material mixed with smooth muscle cells and calcium that develop in the artery walls?
   A. Plaques
   B. Atherosclerosis
   C. Arterial thickening
   D. Cardiovascular disease

2. Tiny, disc-shaped bodies in the blood that are important in clot formation are called __________
   A. plaques.
   B. T-cells.
   C. platelets.
   D. thrombocytes.

3. What should be the first action taken to lower blood cholesterol?
   A. Begin drug treatment.
   B. Consume a high protein diet.
   C. Consume large amounts of fish and fish oils.
   D. Achieve and maintain appropriate body weight.

4. Which of the following defines the association between nutrition and chronic disease?
   A. Diet can influence the time of onset of some chronic diseases.
   B. Diet is the primary factor affecting the development of chronic diseases.
   C. Dietary influence in the development of chronic diseases is direct, straightforward, and well understood.
   D. Dietary advice for combating heart disease and cancer prevents their development if instituted early in life.

5. How does obesity increase the risk for development of hypertension?
   A. The excess fat pads surrounding the kidneys impair blood flow to these organs and lead to higher output of renin.
   B. Sodium intake in the obese significantly exceeds the recommended intake, thereby predisposing to higher blood pressure.
   C. Higher activities of lipoprotein lipase in the obese triggers the angiotensin cascade leading to increased peripheral resistance to blood flow.
   D. Obesity frequently is associated with high blood insulin levels which signal the kidneys to retain sodium, which increases blood pressure.

6. Diuretics act to lower blood pressure by __________
   A. increasing fluid loss.
   B. decreasing potassium loss.
   C. reducing arterial plaque formation.
   D. increasing retention of calcium and potassium.

7. High blood pressure or hypertension is caused by __________
   A. obesity.
   B. smoking.
   C. lack of exercise.
   D. all of the above.

8. On the average and compared to men, women tend to get heart disease __________
   A. less frequently than men.
   B. later in life than men.
   C. with fewer complications than men.
   D. all of the above.

9. Populations that may be more susceptible to potential heart damage caused by elevated triglycerides include people with __________
   A. osteoporosis.
   B. diabetes.
   C. cancer.
   D. AIDS.

10. When LDL level is elevated and HDL is lower than recommended, the first thing that should be examined is a person's __________
    A. alcohol intake.
    B. level of exercise.
    C. genetic background.
    D. smoking habits.

11. People at very high risk for cardiovascular disease should keep their LDL cholesterol levels below ________
    A. 200.
    B. 130.
    C. 100.
    D. 70.

Matching: Select the letter next to the word or phrase that best matches the numbered statements. Answers are used only once.

_____ 12. The primary cause of death of Americans

_____ 13. A mound of lipid material embedded in arterial walls

_____ 14. A blood clot that is attached to arterial plaque

A. Cancer
B. Plaque
C. Embolism
D. Stroke
E. Cardiovascular disease

Essay: Reflect on the item(s) listed below; one or more may be included on the exam.

15. Outline the recommendations for reducing the risk of cardiovascular disease.

16. Describe the major recommendations for reducing the risk for hypertension.

ANSWER KEY

The following provides the answers and references for the Practice Test questions.

	Answer	Learning Objective	Reference
1.	A	LO 2, 3	textbook, p. 588
2.	C	LO 2	textbook, p. 588
3.	D	LO 4, 5	textbook, p. 590
4.	A	LO 1	textbook, p. 585
5.	D	LO 6	textbook, p. 595
6.	A	LO 7	textbook, p. 597
7.	D	LO 1, 6	Video
8.	B	LO 4	Video
9.	B	LO 4	Video
10.	B	LO 5	Video
11.	D	LO 5, 8	Video
12.	E	LO 4	textbook, p. 586
13.	B	LO 2	textbook, pp. 586–587
14.	C	LO 3	textbook, p. 587
15.		LO 5	textbook, pp. 593–594
16.		LO 8	textbook, pp. 596–597

Lesson 23

Diet and Health: Cancer, Immunology, and AIDS

LESSON ASSIGNMENTS

Text: Whitney and Rolfes, *Understanding Nutrition*, Chapter 18, "Diet and Health," pp. 581–586 and pp. 602–609

Video: "Diet and Health: Cancer, Immunology, and AIDS" from the series *Nutrition Pathways*

Optional Web Activities:
Consult your instructor and/or syllabus for any assigned activities.

OVERVIEW

Cancer—the most dreaded of all diseases—is the second leading cause of death in the United States. Some cancers can be cured if detected in time; others are incurable; still others can be prevented altogether if proper changes in lifestyle and eating habits are made in a timely manner. This lesson provides you with the information necessary to take action against the onset of cancer and includes an explanation of how cancer develops, the risk factors associated with cancer, and the recommendations for reducing the risk of cancer. In addition to learning about cancer, you will also learn about the basic functioning of the immune system, the impact of nutrition and lifestyle on immunity, and how nutrition intervention can help alleviate the symptoms and improve the outcome of HIV/AIDS patients.

LEARNING OBJECTIVES

Upon completing this lesson, you should be able to:

1. Describe various cancers as classified by tissue or cells from which they develop.
2. Describe how genetics, environment, and diet can play a role in the development of cancer.
3. State the *Diet and Health* recommendations for reducing cancer risk.
4. Describe the impact of nutrition on immunity.
5. Define *HIV/AIDS*, and explain how nutrition intervention can support people with HIV/AIDS.

TEXT FOCUS POINTS

The following focus points are designed to help you get the most from your reading. Review them, then read the assignment. You may want to write notes to reinforce what you have learned.

1. How does cancer rank as a cause of death among people? How do cancers develop? Define the following terms: *cancer*, *carcinogens*, *initiators*, *antipromoters*, *metastasize*, *tumor*, *neoplasm*, *adenoma*, *carcinoma*, *leukemia*, *sarcoma*, *lymphoma*, and *melanoma*.
2. How do genetics influence the development of cancer? What effect does the immune system have on tumor growth? What are environmental factors that are known to cause cancer? What percent of cancers do some researchers think are linked to diet? What dietary factors act as cancer initiators? How do dietary fats impact cancer? What dietary factors promote or antipromote cancer?
3. Cite the guidelines for reducing cancer risk.
4. List the *Diet and Health* recommendations to prevent or forestall chronic diseases. How should individuals determine whether dietary recommendations for chronic diseases are important to them?
5. Define *immune system*, *phagocytes*, and *lymphocytes*. What are the body's first lines of defense against foreign substances? Differentiate between B-cells and T-cells. How does malnutrition impact the immune system? Briefly describe how AIDS develops. What is "synergistic"?

VIDEO FOCUS POINTS

The following focus points are designed to help you get the most from the video segment of this lesson. Review them before watching the video. You may want to write notes to reinforce what you have learned.

1. What is the impact of fat on the development of breast cancer? What is the recommended fat intake to prevent breast cancer? What are the six suggestions from the American Cancer Society regarding cancer prevention? Which type of fiber has an impact on cancer development?

2. How does immune function affect nutritional status of hospitalized people? Why do experts study the elderly with regard to immune function? What are simple rules to follow regarding strengthening the immune system? What do experts advise with regard to antioxidant supplementation and immune system function and health?

3. Define *HIV* and *AIDS*. What impact does malnutrition have on AIDS? What nutrition strategies are provided to AIDS patients? Describe how nontraditional supplements might impact AIDS. What are the dietary guidelines for people with AIDS? What are the most commonly malabsorbed nutrients in AIDS patients? What is the recommendation concerning vitamin supplements for AIDS patients? Describe the impact of anorexia on AIDS patients.

PRACTICE TEST

The following items will help you check your understanding of this lesson. Compare your answers to the Answer Key at the end of the lesson. Review the course materials related to any incorrect answer.

Multiple Choice: Select the one choice that best answers the question.

1. A cancer that originates from bone is a/an __________
   A. sarcoma.
   B. carcinoma.
   C. osteocarcinoma.
   D. hematopoietic neoplasm.

2. Which of the following statements represents current thought in the development of cancer?
   A. Fat in the diet appears to be protective against many types of cancer.
   B. Food additives play only a small role, if any, in the causation of cancer.
   C. Food contaminants play only a small role, if any, in the causation of cancer.
   D. Protein in the diet from animal sources appears to be protective against many types of cancer.

3. Antioxidant substances that are believed to help protect against cancer include all of the following EXCEPT __________
   A. EDTA.
   B. vitamin E.
   C. vitamin C.
   D. beta-carotene.

4. Which of the following is NOT among the recommendations issued by health professionals to reduce cancer risks?
   A. Moderate or stop intake of alcohol.
   B. Increase intake of foods high in iron.
   C. Limit intake of salted foods and smoked meats.
   D. Eat at least seven servings a day of whole grains, starch, vegetables, and legumes.

5. Immunoglobulins are produced primarily by __________
   A. T-cells.
   B. B-cells.
   C. antigens.
   D. phagocytes.

6. Experts suggest that the threshold of fat intake to prevent breast cancer is __________
   A. less than 30 percent of total kcalories.
   B. less than 20 percent of total kcalories.
   C. about 30 percent of total kcalories.
   D. about 45 percent of total kcalories.

7. The American Cancer Society suggests that women and men maintain desirable weight based on the following criteria:
   A. women should weigh 100 pounds for 5 feet plus 5 pounds for each inch over 5 feet.
   B. both women and men should weigh 106 pounds for 5 feet plus 6 pounds for each inch over 5 feet.
   C. men should weigh 106 pounds for 5 feet plus 6 pounds for each inch over 5 feet.
   D. A and C.

8. Among hospitalized patients, health professionals see a general pattern with regard to immune function in that __________
   A. appetite decreases and patients become malnourished.
   B. immune function improves because of overall good care.
   C. lack of nutrients leads to faulty immune function.
   D. A and C.

9. By studying the immune systems of elderly people, health professionals are learning __________
   A. what nutrients appear promising in reducing risk of diseases such as cancer.
   B. that nutrition helps people to grow old more gracefully.
   C. to turn back the hands of time by supplementing with specific nutrients.
   D. all of the above.

10. HIV can be defined as __________
    A. a virus.
    B. an infection that attacks the immune system.
    C. a disease that attacks the T-4 cells in the body.
    D. all of the above.

11. The most commonly malabsorbed nutrient(s) in AIDS patients is(are) __________
    A. lactose or milk sugar.
    B. protein or amino acids.
    C. fatty acids.
    D. A and C.

12. Very high protein intakes are recommended to AIDS patients because they __________
    A. become anorexic and lose lean body mass.
    B. are in negative nitrogen balance due to wasting.
    C. cannot tolerate high carbohydrate foods as a rule.
    D. A and B.

Matching: Select the letter next to the word or phrase that best matches the numbered statements. Answers are used only once.

______	13. A substance or event that gives rise to a cancer	A. Promoter
		B. Inhibitor
______	14. Substance that favors development of a cancer after the cellular DNA has been altered	C. Initiator

Essay: Reflect on the item(s) listed below; one or more may be included on the exam.

15. Outline the steps involved in the development of cancer.

16. A. How does AIDS develop?
    B. Describe the wasting syndrome of HIV.

ANSWER KEY

The following provides the answers and references for the Practice Test questions.

	Answer	Learning Objective	Reference
1.	A	LO 1	textbook, p. 603
2.	B	LO 2	textbook, p. 604
3.	A	LO 3	textbook, p. 585
4.	B	LO 3	textbook, pp. 605–606
5.	B	LO 4	textbook, p. 582
6.	B	LO 3	Video
7.	D	LO 2, 3	Video
8.	D	LO 4	Video
9.	A	LO 4	Video
10.	D	LO 5	Video
11.	D	LO 5	Video
12.	D	LO 5	Video
13.	C	LO 1	textbook, p. 602
14.	A	LO 2	textbook, p. 602
15.		LO 2	textbook, pp. 602–605
16.		LO 5	textbook, p. 584

Lesson 24

Diet and Health: Diabetes

LESSON ASSIGNMENTS

Text: Whitney and Rolfes, *Understanding Nutrition*, Chapter 18, "Diet and Health," pp. 597–602

Video: "Diet and Health: Diabetes" from the series *Nutrition Pathways*

Optional Web Activities:
Consult your instructor and/or syllabus for any assigned activities.

OVERVIEW

The sixth leading cause of death in the United States is diabetes. Not only is it among the top ten causes of death, but people with diabetes are also two times as likely to die from cardiovascular conditions such as heart attack and stroke as people without the disease. In this lesson you will explore the differences between the two most common forms of diabetes mellitus—insulin-dependent diabetes mellitus (type 1) and noninsulin-dependent diabetes mellitus (type 2). Specifically, you will learn which form is most common, what complications are associated with each form, what the dietary recommendations are for type 1 and type 2, and how exercise impacts the diabetic condition.

LEARNING OBJECTIVES

Upon completing this lesson, you should be able to:

1. Define *diabetes mellitus* and related terms.

2. Differentiate between type 1 and type 2.

3. Explain the complications associated with diabetes.

4. Describe the dietary recommendations for diabetes.

5. Explain how exercise impacts diabetes.
6. Explain how an alternative treatment might impact diabetes.

TEXT FOCUS POINTS

The following focus points are designed to help you get the most from your reading. Review them, then read the assignment. You may want to write notes to reinforce what you have learned.

1. Define the following terms: *diabetes mellitus*, *prediabetes, type 1 diabetes*, *type 2 diabetes*, and *insulin resistance.*

2. What percent of the population in the United States has type 1 diabetes? What is the function of insulin? Why must insulin be injected rather than taken orally? What percent of the population in the United States has type 2 diabetes? In what population does type 2 diabetes most often develop? What percent of U.S. adults with type 2 diabetes are obese? What happens to insulin receptors on cells as body fat increases?

3. Describe the complications associated with diabetes. Approximately what percent of people with diabetes die from CVD? Approximately what percent of people with diabetes have kidney or vision impairment? What is gangrene, and what can happen as a result of gangrene?

4. What is the focus of nutrition therapy in diabetes? Under what three circumstances must a person with type 1 diabetes adjust insulin dosage? Describe the dietary recommendations for type 2 diabetes. What is the effect of moderate weight loss on type 2 diabetes?

5. What is the impact of physical activity on insulin and blood glucose?

VIDEO FOCUS POINTS

The following focus points are designed to help you get the most from the video segment of this lesson. Review them before watching the video. You may want to write notes to reinforce what you have learned.

1. Why must people with type 1 diabetes take insulin injections? Describe the acute symptoms associated with type 1 diabetes.

2. What are the difficulties associated with diagnosis of type 2 diabetes? What population is the highest risk group for type 2 diabetes?

3. What is the primary role for diabetes educators with regard to parents of children with type 1 diabetes? Why should children with type 1 diabetes be monitored carefully? What dietary restrictions do children with type 1 diabetes have? What is the benefit of exercise for children with type 1 diabetes?

4. What four factors are associated with the onset of type 2 diabetes? What is the number one condition associated with type 2 diabetes? What percent of nontraumatic, nonaccidental amputations is associated with type 2 diabetes? What percent of new cases of renal failure are attributed to type 2 diabetes? What percent of people with type 2 diabetes will die from cardiovascular disease?

5. Why is exercise encouraged early on in the management of type 2 diabetes?

6. What are the effects of a raw vegetable diet on type 2 diabetes?

PRACTICE TEST

The following items will help you check your understanding of this lesson. Compare your answers to the Answer Key at the end of the lesson. Review the course materials related to any incorrect answer.

Multiple Choice: Select the one choice that best answers the question.

1. In which of the following conditions would the pancreas be unable to synthesize the hormone insulin?
   A. Hyperglycemia
   B. Type 1 diabetes mellitus
   C. Type 2 diabetes mellitus
   D. Adult-onset diabetes mellitus

2. A person with diabetes is most likely to develop __________
   A. AIDS.
   B. cancer.
   C. diverticulosis.
   D. strokes and heart attacks.

3. Gangrene is a common complication in people with __________
   A. cancer.
   B. diabetes.
   C. pancreatitis.
   D. HIV infection.

4. Insulin injections are required daily for people with type 1 diabetes because __________
   A. their bodies do not produce insulin.
   B. it prevents many metabolic problems from occurring.
   C. insulin stimulates the cells to remove glucose from the blood.
   D. all of the above.

5. Because onset is more gradual with type 2 diabetes, making diagnosis difficult, experts surmise that __________
   A. more than one-fourth of all people with diabetes remain undiagnosed.
   B. more than one-third of all people with diabetes remain undiagnosed.
   C. more than one-half of all people with diabetes remain undiagnosed.
   D. more than three-fourths of all people with diabetes remain undiagnosed.

6. The population at highest risk for type 2 diabetes is __________
   A. males under forty years.
   B. females over forty years.
   C. children under eleven years.
   D. adolescents between fourteen and eighteen years.

7. The primary role of diabetes educators when children with type 1 diabetes are released from the hospital is to __________
   A. teach parents and families how to give insulin injections.
   B. give the children their insulin injections until they are old enough to give them to themselves.
   C. teach parents how to perform blood glucose tests.
   D. A and C.

8. Care must be taken with regard to exercise in children with type 1 diabetes because exercise __________
   A. can decrease the amount of insulin needed during the day.
   B. has been shown to increase the amount of insulin needed.
   C. is not beneficial to children with type 1 diabetes.
   D. A and C.

9. Of which of the following conditions/diseases is diabetes the number one cause?
   A. Heart disease
   B. Blindness
   C. Amputation
   D. Kidney failure

10. Glucose uptake can be increased by 10 to 20 percent above normal if a person with diabetes __________
    A. takes oral agents in addition to insulin injections.
    B. eats only foods high in complex carbohydrates and animal protein.
    C. exercises on a regular basis.
    D. all of the above.

11. Exercise is encouraged early on in the treatment of diabetes because exercise can __________
    A. cause muscles to take up ten to twenty times more glucose than nonexercising muscles.
    B. prevent the heart problems associated with diabetes.
    C. prevent amputations because exercising legs are stimulated to release stored glycogen.
    D. burn excess kcalories and reduce body fat associated with the onset of diabetes.

Matching: Select the letter next to the word or phrase that best matches the numbered statements. Answers are used only once.

_____ 12. This type of diabetes is usually controlled without insulin injections

_____ 13. Term that describes the death of tissue due to deficient blood supply, common in severe diabetic cases

A. Gangrene
B. Polydipsia
C. Type 1 diabetes
D. Type 2 diabetes

Essay: Reflect on the item(s) listed below; one or more may be included on the exam.

14. Compare and contrast the two major types of diabetes and their recommended dietary management.

ANSWER KEY

The following provides the answers and references for the Practice Test questions.

	Answer	Learning Objective	Reference
1.	B	LO 1, 2	textbook, p. 598
2.	D	LO 3	textbook, p. 599
3.	B	LO 3	textbook, p. 601
4.	D	LO 2, 6	Video
5.	C	LO 2	Video
6.	B	LO 2	Video
7.	D	LO 3	Video
8.	A	LO 5	Video
9.	B	LO 3	Video
10.	C	LO 5, 6	Video
11.	A	LO 5	Video
12.	D	LO 2	textbook, pp. 598–599
13.	A	LO 3	textbook, p. 601
14.		LO 1, 2, 3, 4	textbook, pp. 601–602

Lesson 25

Consumer Concerns and Food Safety

LESSON ASSIGNMENTS

Text: Whitney and Rolfes, *Understanding Nutrition*, Chapter 19, "Consumer Concerns about Foods and Water," pp. 623–656

Video: "Consumer Concerns and Food Safety" from the series *Nutrition Pathways*

Optional Web Activities:
Consult your instructor and/or syllabus for any assigned activities.

OVERVIEW

As consumers we are all concerned about what goes into the foods we eat and the water we drink before they get to our tables. How do pesticides and additives affect the foods we eat and our health? Should we worry about the public water supply and drink only bottled water? These questions are examined and answers are offered to you in Lesson 25. Specifically, this lesson introduces you to food-borne illnesses and explains how you can prevent them by practicing food safety in your home and while traveling. In addition, this lesson explains how environmental contaminants make their way into the food and water supplies, what the functions and hazards associated with pesticide use are, how food additives affect foods and consumer health, how the public water supply may be contaminated and how it is protected, and how bottled water issues impact consumers.

LEARNING OBJECTIVES

Upon completing this lesson, you should be able to:

1. Define terms associated with food-borne illnesses. Explain how food-borne illnesses impact health.
2. Explain how to prevent food-borne illness through food safety in the kitchen and while traveling.
3. Describe in general terms how environmental contaminants find their way into the food supply.
4. Explain the function of pesticides, the hazards associated with pesticides, and ways to minimize risks. Identify pesticide alternatives.
5. Describe how food additives are used in the food supply, including how food additives may affect health.
6. Describe the sources of drinking water, including how contamination is prevented.
7. Explain how bottled water impacts the consumer.

TEXT FOCUS POINTS

The following focus points are designed to help you get the most from your reading. Review them, then read the assignment. You may want to write notes to reinforce what you have learned.

1. Define the following terms: *food-borne illness*, *food-borne infection*, and *food intoxication*. How many cases of food-borne illnesses are reported in the United States each year? What populations are most vulnerable to food-borne illnesses? Name the two most common food-borne infectious agents, the foods that house them, and their symptoms. What is the most common food toxin?
2. What percent of reported food-borne illnesses occur in commercial settings? Explain the purpose of the Hazard Analysis Critical Control Points (HACCP) system. What four steps can be taken to prevent food-borne illnesses from arising in the kitchen? Why should meat be cooked thoroughly? How should

meat and poultry be thawed? What is a general rule concerning a suspect food? What percent of people who travel to other countries contract travelers' diarrhea? Cite ways to avoid food-borne illness while traveling. What is irradiation, and what are the consumer concerns regarding the process?

3. Define the following terms: *contaminant*, *methylmercury*, and *polychlorinated biphenyl* (PCB) and *polybrominated biphenyl* (PBB). What is the agency that regulates commercial fishing to help ensure safety standards are met?

4. What is solanine? Define *pesticide*. What populations are at risk for pesticide hazards? Briefly describe how the EPA and FDA monitors pesticide use. How can consumers minimize risks associated with pesticide use? What are alternatives to pesticide use? If farmers want to produce and market "organically grown" crops, what are the USDA regulations that must be applied?

5. What is the general function of additives in the food supply? What three steps must be taken by food manufacturers to satisfy the FDA regarding additives? Describe the function of the GRAS (generally recognized as safe) list and the Delaney Clause. What are four functions for which additives should not be used according to the FDA? Explain how the following intentional food additives affect food and/or health: antimicrobial agents, nitrites, antioxidants, sulfites, BHA/BHT, artificial colors, flavor enhancers, and sugar alternatives, texture and stability, and nutrient additives. List indirect food additives, and explain how they may impact health.

6. What are the sources of drinking water for the population? What is the source of surface water? How does surface water become contaminated? What is the source of groundwater? How does groundwater become contaminated?

7. How is bottled water classified by the FDA? What percent of bottled water comes from groundwater? What is the disinfectant used in bottled water?

VIDEO FOCUS POINTS

The following focus points are designed to help you get the most from the video segment of this lesson. Review them before watching the video. You may want to write notes to reinforce what you have learned.

1. What is a common belief regarding bottled water versus tap water? How do bottled water sales compare to other beverages? How can consumers determine the source of bottled water?

2. What are alternate methods of growing herbs, without use of pesticides, that also enhance the quality of the soil? What types of plants or insects may be used to repel harmful plant pests?

3. What are some of the side effects of monosodium glutamate (MSG) as a food additive? What can people do to avoid MSG?

4. What are the three components of safe food handling? What is cross-contamination? What advice is given when handling fruits and vegetables during preparation?

PRACTICE TEST

The following items will help you check your understanding of this lesson. Compare your answers to the Answer Key at the end of the lesson. Review the course materials related to any incorrect answer.

Multiple Choice: Select the one choice that best answers the question.

1. Which of the following is the major food source for transmission of *Campylobacter jejuni*?
   A. Raw poultry
   B. Uncooked seafood
   C. Contaminated water
   D. Imported soft cheeses

2. Among the following organisms, which is primarily responsible for causing travelers' diarrhea?
   A. *Vibrio*
   B. *Escherichia coli*
   C. *Clostridium botulinum*
   D. *Staphylococcus aureus*

3. What organism produces a deadly toxin in anaerobic conditions such as improperly canned goods?
   A. *Salmonella*
   B. *Clostridium botulinum*
   C. *Staphylococcus aureus*
   D. *Listeria monocytogenes*

4. Which of the following would most likely result from placing cooked hamburger patties on the same plate that held the uncooked patties?
   A. Flavor declination
   B. Meat juice retention
   C. Fat drippings exudation
   D. Microbial cross-contamination

5. Which of the following is a feature of certain food treatments or additives?
   A. Irradiation is a form of microwaves.
   B. Sulfites act to retard growth of pathogenic organisms.
   C. Carotenoids are used to retard formation of nitrosamines.
   D. Nitrites may form unique radiolytic particles when the food is overheated.

6. All of the following practices are known to minimize exposure to food pesticide residues EXCEPT __________
   A. throwing away the outer leaves of leafy vegetables.
   B. using a knife to peel citrus fruits rather than biting into the peel.
   C. throwing away the fats and oils in broths and pan drippings from cooked meats.
   D. washing waxed fruits and vegetables in water to remove the wax-impregnated pesticides.

7. Of the following, which is used most widely as an antimicrobial agent?
   A. Sugar
   B. Saccharin
   C. Sodium nitrite
   D. Sodium propionate

8. When a slice of fresh apple turns a brown color it is most likely the result of __________
   A. oxidation.
   B. dehydration.
   C. microbial contamination.
   D. ethylene oxide treatment in the ripening process.

9. What substance is commonly added to public water supplies to disinfect the water?
   A. Ozone
   B. Fluoride
   C. Chlorine
   D. Penicillin

10. Why have bottled water sales increased dramatically over the past several years?
    A. People believe that tap water from the public water supply is unhealthy or unclean.
    B. Manufacturers of bottled water have made it inexpensive to buy when compared to tap water.
    C. Grocery stores have increased the shelf space, making bottled water more visible to consumers.
    D. All of the above.

11. An example of an insect that can be used to repel or eat other insects that damage or destroy crops is the __________
    A. cockroach.
    B. June bug.
    C. ladybug.
    D. earthworm.

12. In order to avoid the side effects of MSG or other food additives, people should __________
    A. learn to read labels to find the offensive additive.
    B. eat more fresh fruit, vegetables, and whole grains.
    C. ask that the offensive food additive be left out of foods when eating out.
    D. all of the above.

13. Time and temperature are two components of safe food handling. The third basic component is __________
    A. speed.
    B. cleanliness.
    C. frequency.
    D. intensity.

14. When handling fruits, vegetables, and meats in the kitchen, the best advice is to __________
    A. avoid cross-contamination by sanitizing utensils.
    B. keep foods refrigerated until ready to eat, then serve immediately.
    C. avoid taking a bite of raw food unless it's stamped "Safe for Consumption."
    D. wash hands and face immediately after handling meats.

Fill in the Blank: Insert the correct word or words in the blank for each item.

15. Sales of soft drinks, beer, and wine have taken a back seat compared to sales of ____________________.

16. One alternative to pesticide use in organic farming is ________________.

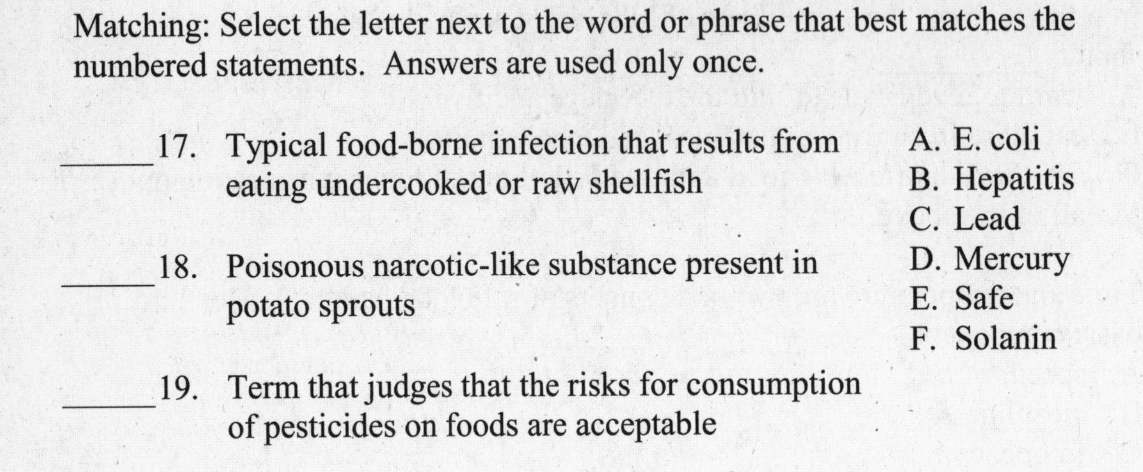

Matching: Select the letter next to the word or phrase that best matches the numbered statements. Answers are used only once.

_____17. Typical food-borne infection that results from eating undercooked or raw shellfish

_____18. Poisonous narcotic-like substance present in potato sprouts

_____19. Term that judges that the risks for consumption of pesticides on foods are acceptable

A. E. coli
B. Hepatitis
C. Lead
D. Mercury
E. Safe
F. Solanin

Essay: Reflect on the item(s) listed below; one or more may be included on the exam.

20. List three major pathogenic microbes that are transmitted by foods. Describe their food sources, symptoms of sickness, and methods of prevention.

21. List the major antioxidant additives in the food supply and their side effects in human beings.

ANSWER KEY

The following provides the answers and references for the Practice Test questions.

	Answer	Learning Objective	Reference
1.	A	LO 1	textbook, p. 625
2.	B	LO 1, 2	textbook, p. 625
3.	B	LO 1	textbook, p. 625
4.	D	LO 3	textbook, p. 628
5.	A	LO 5	textbook, p. 632
6.	D	LO 4	textbook, p. 628
7.	A	LO 4	textbook, pp. 643–644
8.	A	LO 5	textbook, p. 634
9.	C	LO 6	textbook, p. 643
10.	A	LO 7	Video
11.	C	LO 4	Video
12.	D	LO 5	Video
13.	B	LO 2	Video
14.	A	LO 2	Video
15.	bottled water	LO 7	Video
16.	Answer may include any of the following: Earthworm castings Bat guano Seedless cow manure Ladybugs Compatible plants	LO 4	Video
17.	B	LO 1	textbook, p. 631
18.	F	LO 3	textbook, p. 637
19.	E	LO 4	textbook, p. 626
20.		LO 1, 2, 3	textbook, pp. 625–626
21.		LO 5	textbook, pp. 644–645

Lesson 26

Applied Nutrition

LESSON ASSIGNMENTS

Text: There is no textbook reading assignment for this lesson.

Video: "Applied Nutrition" from the series *Nutrition Pathways*

Optional Web Activities:
Consult your instructor and/or syllabus for any assigned activities.

OVERVIEW

This lesson concludes the twelve-month travels of the three *Pathways* subjects profiled throughout this video-based course. You observed them as they were evaluated by nutrition or other allied health professionals; you watched them as they traveled down their respective pathways to healthier lives; and you examined the choices they made and the challenges they faced along the way. In this final lesson, you will see them interact with each other and with the professionals that assisted them in their journey in a talk show format hosted by Debra Duncan of the ABC Network affiliate in Dallas, Texas. The host asks the subjects to recap their experiences and to explain the difficulties they faced and the successes they had as they attempted to change their lives for the better. To conclude, each subject offers advice to help others achieve success in implementing personal nutrition and lifestyle changes to improve health and well-being.

LEARNING OBJECTIVES

Upon completing this lesson, you should be able to:

1. Explain why the three *Pathways* subjects chose to participate in the program.
2. Explain the reasons why some people fail when attempting to change nutrition and lifestyle habits.
3. Describe the most significant outcomes of the program for each subject.
4. Summarize the nutrition philosophy associated with this program.
5. State the importance of exercise in achieving success with regard to improvements in health and well-being.
6. State the advice you would give to someone who wants or needs to change nutrition or lifestyle habits.

VIDEO FOCUS POINTS

The Video Focus Points for this lesson are identical to the above Learning Objectives.

PRACTICE TEST

The following items will help you check your understanding of this lesson. Compare your answers to the Answer Key at the end of the lesson. Review the course materials related to any incorrect answer.

Multiple Choice: Select the one choice that best answers the question.

1. The reasons the *Pathways* subjects chose to participate in the *Nutrition Pathways* program included all of the following EXCEPT __________
   A. a family history of heart disease.
   B. a family history of diabetes.
   C. having colon cancer.
   D. being overweight.

2. The goal of the weight-loss subject to lose forty pounds in one year was for her __________
   A. realistic.
   B. unrealistic.
   C. achieved.
   D. B and C.

3. Among the reasons why people fail in changing nutrition and lifestyle habits include __________
   A. setting goals too high.
   B. changing old habits too slowly.
   C. expecting too little from the efforts that go into changing old habits.
   D. all of the above.

4. The greatest success for the subject who has type 2 diabetes involved __________
   A. losing fifty pounds by the end of the year.
   B. jogging two miles in twenty minutes.
   C. going from two insulin injections per day to none.
   D. all of the above.

5. The philosophical emphasis throughout *Nutrition Pathways* can be best summed up in which one of the following statements?
   A. "Change your life."
   B. "Balance your choices."
   C. "Deprivation breeds success."
   D. "Variety equals challenge."

Essay: Reflect on the item(s) listed below; one or more may be included on the exam.

6. Describe the importance of exercise with respect to stress reduction and balance in people's lives and its impact on type 2 diabetes, high blood cholesterol, and overweight.

7. What advice would you give to someone who wanted or needed to change nutrition or lifestyle habits? Include comments on the following: the need for evaluation of current physical and/or nutrition status, steps necessary to improve nutrient status or physical status, ways to handle challenges that might

interfere with healthful choices, and the importance of balance and moderation in choices.

ANSWER KEY

The following provides the answers and references for the Practice Test questions.

Answer	Learning Objective	Reference
1. C	LO 1	Video
2. B	LO 2	Video
3. A	LO 2	Video
4. C	LO 3	Video
5. B	LO 4	Video
6.	LO 5	Video
7.	LO 6	Video

Contributors

We gratefully acknowledge the valuable contributions to this course from the following individuals. The titles listed were accurate when the video programs were recorded, but may have changed since the original taping.

LESSON 1—"NUTRITION BASICS"

Noemi Cruz, R.D., Registered Dietitian, Hartford Hospital, Hartford, CT
Jeanne P. Goldberg, Ph.D., Director, Center on Nutrition Communication, Tufts University, Boston, MA
Ann Zogbaum, M.S., R.D., Oncology Dietitian, Hartford Hospital, Hartford, CT

LESSON 2—"THE DIGESTIVE SYSTEM"

Judy Suneson, R.D., L.D., Clinical Dietitian, Baylor University Medical Center, Dallas, TX

LESSON 3—"CARBOHYDRATES: SIMPLE AND COMPLEX"

Judith Hallfrisch, Ph.D., Research Leader for the Metabolism and Nutrient Interactions Lab, Beltsville Human Nutrition Research Center, Beltsville, MD

LESSON 4—"CARBOHYDRATES: FIBER"

James Anderson, Ph.D., Professor of Medicine and Clinical Nutrition, University of Kentucky, Lexington, KY
Mary Sue Cole, R.D., L.D., County Extension Agent, Greenville, TX

LESSON 5—"FATS: THE LIPID FAMILY"

Gloria Vega, Ph.D., Professor of Clinical Nutrition, University of Texas Southwestern Medical Center, Dallas, TX

Clay King, Ph.D., Associate Professor of Nutrition and Food Sciences, Texas Woman's University, Denton, TX

LESSON 6—"FATS: HEALTH EFFECTS"

Jo Ann Carson, M.S., R.D., L.D., Director of Clinical Dietetics Program, University of Texas Southwestern Medical Center, Dallas, TX

LESSON 7—"PROTEIN: FORM AND FUNCTION"

Martha Groblewski, Ph.D., R.D., College of Health Science, Old Dominion University, Norfolk, VA

LESSON 8—"THE PROTEIN CONTINUUM"

Cecile H. Edwards, Ph.D., L.N., C.N.S., Professor of Nutrition, Howard University, Washington, DC

John A. McDougall, M.D., Saint Helena Hospital, Santa Rosa, CA

LESSON 9—"METABOLISM"

Rene Frenkle, Ph.D., Biochemist, University of Texas Southwestern Medical Center, Dallas, TX

Do-Hyun Choe, O.M.D., Doctor of Oriental Medicine, Dallas, TX

Dale Swanholm, M.D., F.A.A.F.P, Board Certified Physician in Family Practice and Geriatrics, Dallas, TX

LESSON 10—"WEIGHT CONTROL: ENERGY REGULATION"

Craig Johnson, Ph.D., Director of Eating Disorders, Laureate Psychiatric Hospital, Clinical Psychology Professor, University of Tulsa, Tulsa, OK
Judith S. Stern, Ph.D., Professor of Nutrition and Internal Medicine, University of California at Davis, Davis, CA

LESSON 11—"WEIGHT CONTROL: HEALTH EFFECTS"

Kelly D. Brownell, Ph.D., Department of Psychology, Yale University, New Haven, CT

LESSON 12—"VITAMINS: WATER SOLUBLE"

Jethon' Sharrieff, Associate Store Director, Whole Foods Market, Chicago, IL
Lillie R. Williams, Ph.D., R.D., Chairperson of the Department of Nutritional Sciences, Howard University, Washington, DC

LESSON 13—"VITAMINS: FAT SOLUBLE"

Ishwarlal Jialal, Ph.D., Associate Professor of Medicine and Pathology, University of Texas Southwestern Medical Center, Dallas, TX

LESSON 14—"MAJOR MINERALS AND WATER"

Claudia A. Barner, Ph.D., R.D., L.D., University of Texas Southwestern Medical School, Dallas, TX
K. Sakhaee, M.D., Professor of Internal Medicine, University of Texas Southwestern Medical Center for Mineral Metabolism and Clinical Research, Dallas, TX

LESSON 15—"TRACE MINERALS"

Rebecca Costello, Ph.D., Deputy Director, Office of Dietary Supplements, National Institutes of Health, Bethesda, MD
Melissa Feagley, R.D., Clinical Dietitian, Plano Presbyterian Hospital, Plano, TX
Barbara Gollman, Ph.D., R.D., Nutrition Consultant, President, Phytopia, Dallas, TX
Jeanette Hasse, Ph.D., R.D., Transplant Nutrition Specialist, Baylor University Medical Center, Dallas, TX
Steve Long, Water System Manager, North Texas Municipal Water Plant, Wylie, TX
Richard Anderson, Ph.D., M.D., Lead Scientist, Beltsville Human Nutrition Research Center, USDA, Beltsville, MD
Jay Shulman, D.D.S., Professor, Baylor College of Dentistry, Dallas, TX
Ann Zogbaum, M.S., R.D., Oncology Dietitian, Hartford Hospital, Hartford, CT

LESSON 16—"PHYSICAL ACTIVITY: FITNESS BASICS"

John Duncan, Ph.D., Chief of Clinical Application, Cooper Institute for Aerobics Research, Dallas, TX

LESSON 17—"PHYSICAL ACTIVITY: BEYOND FITNESS"

Nancy Clark, M.S., R.D., Sports Medicine Brookline, Boston, MA
Melvin H. Williams, Ph.D., Eminent Professor of Health, Physical Education, and Recreation, Old Dominion University, Norfolk, VA

LESSON 18—"LIFE CYCLE: PREGNANCY"

Erica Gunderson, M.S., M.Ed., R.D., Doctoral Program in Epidemiology, University of California at Berkeley, Berkeley, CA

LESSON 19—"LIFE CYCLE: LACTATION AND INFANCY"

Kathryn G. Dewey, Ph.D., Professor in the Department of Nutrition, University of California at Davis, Davis, CA

LESSON 20—"LIFE CYCLE: CHILDHOOD AND ADOLESCENCE"

Craig Johnson, Ph.D., Director of Eating Disorders, Laureate Psychiatric Hospital, Clinical Psychology Professor, University of Tulsa, Tulsa, OK

Laurell Mellin, M.A., R.D., Director of the Center for Child and Adolescent Obesity, Associate Clinical Professor of Family Medicine and Pediatrics, University of California at San Francisco, San Francisco, CA

LESSON 21—"LIFE CYCLE: ADULTHOOD AND AGING"

Floristene Johnson, R.D., Deputy Regional Administrator for the Administration on Aging, Dallas, TX

Robert M. Russell, Ph.D., Associate Director of the Nutrition Research Center on Aging, Professor of Medicine and Nutrition, Tufts University, Boston, MA

LESSON 22—"DIET AND HEALTH: CARDIOVASCULAR DISEASE"

Emily Malorzo, R.D., Registered Dietitian, Baylor Heart and Vascular Hospital, Baylor University Medical Center, Dallas, TX

Ernst J. Schaefer, M.D., Professor of Medicine and Nutrition, Director of Lipid and Heart Disease Prevention Program, Tufts University, Boston, MA

LESSON 23—"DIET AND HEALTH: CANCER, IMMUNOLOGY, AND AIDS"

Stacey J. Bell, D.Sc., R.D., Research Dietitian, New England Deaconess Hospital, Boston, MA
Lauri Wright, R.D., James A. Haley Veterans Hospital, Tampa, FL
Faith D. Ottery, M.D., Ph.D., Surgical Oncologist, Fox Chase Cancer Center, Philadelphia, PA

LESSON 24—"DIET AND HEALTH: DIABETES"

Susan L. Thom, R.D., L.D., C.D.E., 1994–1995 President of American Association of Diabetes Educators, Private Practice, Diabetes Associates, Cleveland, OH
Zaven H. Chakmakjian, M.D., Diplomate of Internal Medicine and Endocrinology/Metabolism, Associate Clinical Professor, University of Texas Southwestern Medical School, Dallas, TX

LESSON 25—"CONSUMER CONCERNS AND FOOD SAFETY"

Michael Jacobson, Ph.D., Founder and Executive Director for the Center for Science in the Public Interest, Washington, DC
Jethon' Sharrieff, Associate Store Director, Whole Foods Market, Chicago, IL

LESSON 26—"APPLIED NUTRITION"

Debbie Childers, R.D., L.D., C.D.E., Diabetes Nutrition Educator, Baylor University Medical Center, Dallas, TX
Donna Israel, Ph.D., R.D., L.D., L.P.C., President of Preferred Nutrition Therapist Incorporated, Owner of the Fitness Formula, Dallas, TX

Diet Analysis Project

ATTENTION: Read the following instructions carefully BEFORE you begin!

Do NOT change your eating habits for this assignment. Your grade will depend on how well you analyze your eating habits, not on the foods you eat! The more carefully you complete your food diary, the more accurate your information will be.

PART I: Keep Daily Food Records and Generate Computer Printouts:

1. Write down all the foods and beverages you eat over a three to seven day period, including ONE weekend day. Record the amount and a description of the food/beverage on the attached Daily Food Record (DFR) form.
   A. Use "ounces, cups, each, teaspoon, tablespoon" etc.
   B. Estimate serving sizes, weigh the food, and/or use food labels.
   C. A three-ounce portion of meat is about the size of a woman's palm or a deck of cards; a restaurant serving of vegetables is ½ to ⅔ cup.
   D. If unsure of serving size, ask the server or estimate on the large side.

2. Input your personal information and your three to seven days of food into the *Diet Analysis Plus* software. Follow the software instructions to input personal data and all foods and beverages.

3. Export your profile and email to your instructor or print the following items:

   A. Your Personal Profile (Daily Recommended Intake based on your gender, activity level, height, weight, and age)

   B. Your three to seven days of food and the AVERAGES including:
      - Ratios and percents
      - Bar graph
      - MyPlate

—Continued on the back

PART II: Answer Questions to Analyze Your Eating Habits:

1. Interpret your results. Refer to the printouts of your three-to-seven-day AVERAGE to obtain data regarding your personal dietary intake. Use the textbook, *Understanding Nutrition*, 13th edition, to answer other nutrition-related questions.

2. Type (or write neatly) in the spaces provided on the question forms. If you need more room to answer questions, use the back of the form or use extra paper.

3. Staple the following pages together and submit them to your professor by the established deadline.
   a. Three-to-seven-day AVERAGE printouts
   b. Questions with answers (Questions are on pp. 321–334 of the Student Course Guide)

NOTE: Students may manually analyze foods by referring to "Appendix H: Table of Food Composition" in the text. Manual analysis is tedious and time consuming, however, and is NOT recommended for distance learning students. Check with your instructor for other software or diet analysis options.

Daily Food Record

NAME: ______________________________ ID #: ___________________________

DIRECTIONS: Write down ALL foods and beverages consumed throughout the day. Be as specific as possible about the amount as well as the description of the food item itself. For example, specify the type of bread, such as whole wheat, white, or rye; type of cheese, such as cheddar, Swiss, or American; whether fruits or vegetables are fresh, frozen, or canned; and if meats are lean only, or lean with some fat. The most accurate entries for amounts are for weighed foods in ounces (oz), grams (gr), or pounds (lb). Other measures are cups (c), teaspoons (tsp), tablespoons (tbs), "each," or "piece." Weights of foods should be for edible portion only.

Try to maintain your *normal* eating patterns. Do NOT exclude foods or beverages that you perceive to be "bad" such as alcohol, candy, etc. You will gain more valuable information about your diet if you honestly analyze your eating habits.

Day #1 (Date: _______________)

AMOUNT:	FOOD AND BEVERAGE DESCRIPTION:

—*Continued on the back*

Day #2 (Date: ____________)

AMOUNT:	FOOD AND BEVERAGE DESCRIPTION:

Day #3 (Date: ____________)

AMOUNT:	FOOD AND BEVERAGE DESCRIPTION:

Carbohydrate and Fiber Intakes

DIRECTIONS: Refer to your three-to-seven-day AVERAGE printouts (generated from the diet software) to answer the following questions. Answer in the space provided.

1. ______________ What is your average percentage of total kcalories from carbohydrates? (See Macronutrient Ranges)

2. ______________ Dietary guidelines state that 45 to 65 percent or more of your total kcalories should come from carbohydrates. How did your intake compare to the recommendation?

3. ______________ It is estimated that you should have 130 grams per day or more of carbohydrate. How many grams per day did you consume on the average? (See Intake vs. Goals Bar Graph)

4. ______________ The recommended intake of fiber per day is 25 to 35 grams. How many grams per day did you consume on the average? (See Intake vs. Goals Bar Graph)

5. ______________ On the average, what percentage of your DRI did you consume in fiber? (See Intake vs. Goals Bar Graph)

6. List three of your most common refined sugar sources (colas, candy, pastries, etc.) that you consume on a regular basis.

   1) ____

   2) ____

   3) ____

7. How many total grams of sugar did you consume? (See Intake vs. Goals Bar Graph.)

—Continued on the back

8. Determine your percentage of calories from sugar: multiply sugar grams by 4 and then divide by your total kcalories. (Show calculations below.)

9. According to some authorities, the recommended intake of sugar should be less than 10 percent of total kcalories. How did your intake compare to the recommendations?

10. List three of your personal best sources of complex carbohydrates or fiber in your diet.

    (1)________________ (2) ________________ (3) ________________

11. List three additional foods you could include in your diet to increase your intake of complex carbohydrates or fiber.

    (1)________________ (2) ________________ (3) ________________

12. How can health be improved by INCREASING your intake of complex carbohydrates and/or fiber? *Cite page numbers in your text to support your answer.*

13. State the advice given to people to prevent complications associated with a sudden increase in fiber intake. Cite page numbers in your text to support your answer.

14. How can health be improved by DECREASING your intake of concentrated sweets? *Cite page numbers in your text to support your answer.*

Fat and Dietary Cholesterol Intakes

DIRECTIONS: Refer to your three-to-seven-day AVERAGE printouts (generated from the diet software) to answer the following questions. Answer in the space provided.

1. ______________ What is your average percentage of total kcalories from fat? (See Macronutrient Ranges.)

2. ______________ The dietary goal for total fat intake is less than 20–35 percent of total kcalories. How did your fat intake compare to the stated goal?

3. ______________ How many total grams of fat did you consume on an average day? (See Intake vs. Goal Bar Graph.)

4. ______________ What percent of your kcalories came from saturated fat? (See Fat Breakdown.)

5. ______________ What percent of your kcalories came from monounsaturated fat?

6. ______________ What percent of your kcalories came from polyunsaturated fat?

7. ______________ How many milligrams of cholesterol did you consume on an average day? (See Intake vs. Goals Bar Graph.)

8. The recommended intake of dietary cholesterol is less than 300 mg per day. How did your intake compare to the recommendation?

—Continued on the back

9. How many grams of the essential fatty acids (EFA) did you consume? (See Intake vs. Goals Bar Graph.)

   (1) Omega-6 (linoleic) ____

   (2) Omega-3 (linolenic) _______________________________________

10. How did your EFA intake compare to the DRI? (See Intake vs. Goals Bar Graph.)

11. List one food containing cholesterol from each day of your three days that you could limit to reduce total cholesterol intake.

    (1)_________________ (2) _________________ (3) _________________

12. List one food containing fat from each day of your three days that you could limit to reduce total fat intake.

    (1)_________________ (2) _________________ (3) _________________

13. How can health be improved by DECREASING total fat, saturated fat, and cholesterol? *Cite page numbers in your text to support answers.*

14. How can health be improved by INCREASING monounsaturated fats, polyunsaturated fats, and omega-3 fats? *Cite page numbers in your text to support answers.*

Protein Intakes

DIRECTIONS: Refer to your three-to-seven-day AVERAGE printouts (generated from the diet software) to answer the following questions. Answer in the space provided.

1. ______________ What is the average percentage of your total kcalories that comes from protein? (See Macronutrient Ranges.)

2. ______________ Dietary guidelines recommend that protein should provide about 10 to 35 percent of the total intake of calories. How did your intake compare to the recommendation?

3. ______________ How many grams of protein did you consume on the average? (See Intake vs. Goals Bar Graph.)

4. ______________ On the average, what percentage of your personal DRI did you consume in protein? (See Intake vs. Goals Bar Graph.)

5. List one MEAT source of protein from each day of your three-day intake. For each meat source of protein, list COMBINED plant sources of protein that you could eat in place of it. (Refer to the Related Activity in Lesson 8, "Plant Proteins for a Meatless Meal," for suggestions on combining plants for high-quality, meatless protein.)

Meat Source:	Plant Combination:
Day 1: ____________________	Day 1: ____________________
Day 2: ____________________	Day 2: ____________________
Day 3: ____________________	Day 3: ____________________

—Continued on the back

For each of the following questions, cite page numbers in your text to support your answer.

6. Explain how eating plant protein versus meat protein impacts the following:

   a. Saturated fat:

   b. Total fat:

   c. Cholesterol:

   d. Fiber:

   e. Total kcalories:

7. How would the use of plant protein in place of meat protein affect the protein quality of the diet?

8. State the health concerns surrounding high-protein diets from animal sources.

Weight Control

DIRECTIONS: Refer to your three-to-seven-day AVERAGE printouts (generated from the diet software) to answer the following questions. Answer in the space provided.

1. ______________ What percentage of your personal DRI did you consume? (See Intake vs. Goals Bar Graph.)

2. ______________ If you consumed the same amount of kcalorie per day over a period of time (see #1 above), would you lose, gain, or maintain your weight?

3. ______________ Is your personal goal to lose, gain, or maintain your weight?

For each of the following questions, cite page numbers in your text to support your answer.

4. What should be the primary consideration when setting a weight goal?

5. How would consuming less than 1,200 kcalories per day impact metabolism?

6. How would consuming less than 1,200 kcalories per day impact health and nutrition status?

7. In addition to changes in your diet, list three other factors that should be included in a successful weight-control program.

   (1)__________________ (2) __________________ (3) __________________

8. What two physical activities could you perform to help you reach and maintain your goal weight?

   (1)______________________________ (2) ______________________________

Vitamin Intakes

DIRECTIONS: Refer to your three-to-seven-day AVERAGE printouts (generated from the diet software) to answer the following questions. Answer in the space provided.

1. Examine your average intake of vitamins. List the water-soluble vitamins that were less than 75 percent of your personal DRI. (See Intake vs. Goals Bar Graph.)

2. List the water-soluble vitamins that were greater than 75 percent of your personal DRI.

3. List the fat-soluble vitamins that were less than 75 percent of your personal DRI.

4. List the fat-soluble vitamins that were greater than 75 percent of your personal DRI.

5. List one good food source for each water-soluble vitamin that was less than 75 percent of your DRI.

6. List one good food source for each fat-soluble vitamin that was less than 75 percent of your DRI.

—Continued on the back

For each of the following questions, cite page numbers in your text to support your answer.

7. Explain why excess water-soluble vitamin intakes are less of a health concern than excess fat-soluble vitamin intakes.

8. State the health concerns surrounding vitamin supplements in general.

9. State the health concerns surrounding antioxidant vitamins specifically.

Mineral Intakes

DIRECTIONS: Refer to your three-to-seven-day AVERAGE printouts (generated from the diet software) to answer the following questions. Answer in the space provided.

1. Examine your average intake of minerals. List the major minerals that were less than 75 percent of your personal DRI. (See Intake vs. Goals Bar Graph)

2. List the major minerals that were greater than 75 percent of your personal DRI.

3. List the trace minerals that were less than 75 percent of your personal DRI.

4. List the trace minerals that were greater than 75 percent of your personal DRI.

5. List one good food source for each major mineral that was less than 75 percent of your DRI.

6. List one good food source for each trace mineral that was less than 75 percent of your DRI.

—Continued on the back

For each of the following questions, cite page numbers in your text to support your answer.

7. Health recommendations state that sodium intake should be limited to 2,300 mg per day.

   a. How did your intake of sodium compare to the recommendation?

   b. How could more than 2,300 mg per day of sodium adversely affect health?

8. State the health concerns surrounding iron overload.

The MyPlate and Other Information

DIRECTIONS: Refer to your three-to-seven-day AVERAGE printouts (generated from the diet software) to answer the following questions. Answer in the space provided.

THE MyPlate: Refer to *MyPlate Analysis* to support your answers.

1. Cite your personal Goal and % Goal for each of the following groups on your Pyramid.

	Goal	% Goal
Grains	____________________	____________________
Vegetables	____________________	____________________
Fruits	____________________	____________________
Dairy	____________________	____________________
Protein Foods	____________________	____________________
Empty Calories	____________________	____________________

2. What is the recommended number of kcalories for your MyPlate?

3. Provide a brief explanation of "empty calories."

4. Give two examples of how you can use empty calorie allowance.
   a.

   b.

—*Continued on the back*

5. According to the www.choosemyplate.gov website, what are the recommendations regarding physical activity for health benefits?

6. According to the www.choosemyplate.gov website, cite four benefits for physical activity.